Muhammetnur Atayev
Palvan Myradov
Jumamuhamet Abdyyev

Software Revolution: Новые горизонты программирования

Muhammetnur Atayev
Palvan Myradov
Jumamuhamet Abdyyev

Software Revolution: Новые горизонты программирования

Collection of scientific articles

LAP LAMBERT Academic Publishing

Cover image: www.ingimage.com

Publisher:
LAP LAMBERT Academic Publishing
is a trademark of
Dodo Books Indian Ocean Ltd. and OmniScriptum S.R.L publishing group

120 High Road, East Finchley, London, N2 9ED, United Kingdom
Str. Armeneasca 28/1, office 1, Chisinau MD-2012, Republic of Moldova, Europe
Managing Directors: Ieva Konstantinova, Victoria Ursu
info@omniscriptum.com

Printed at: see last page
ISBN: 978-620-9-25127-6

CONTENT

UDC 004.4

TO CREATE A PROGRAM SUPPLY FOR REGISTRATION

M.A. Atayev, P.S. Myradov, J.R. Abdyyev
The State Energy Institute of Turkmenistan, Mary, Turkmenistan

Abstract. Based on this program, customers (sellers, storekeepers, store managers) can see the list, prices and characteristics of drugs which available in the pharmacy at any time. This will help save a lot of time for the users of this application. This program can be installed and used easy in any pharmacy.

This program is written by Delphi programming language. When the main window of the program opens, a simple and intuitive interface will appear in front of the user. The list of medications on the left side of the home page,

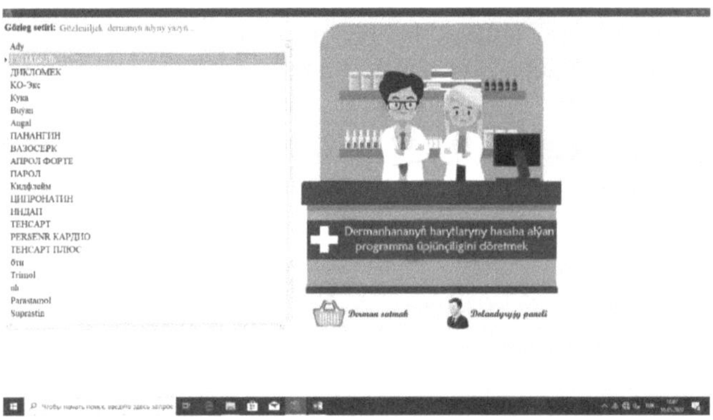

Figure 1. Application interface

It helps as an auxiliary tool for the user. That is, when you click on any drug in the list information about the drug will appear in a separate window. In that window, written information about what diseases the drug is used for, its manufacturer, etc.

Also, there are 2 buttons took place in the lower left corner of the main window for working in the program:

-Selling medicine

-Control panel

When you click on the “Sell Medicines” button, there will appear a window as following. At first the name of the drug to be sold must be entered in this window. Then you need to enter its amount and click on the button called

"Add to List". There is also an option to print this transaction after entering all the drugs to be sold, for this must be clicked the "Print" button.
The interface of task window in Figure 1 and code written for the required information in the Delphi programming language are as follows:

```
procedure TForm1.Edit1Change(Sender: TObject);
var help1, help2:string;
begin
help1:='%'+Edit1.Text+'%';
help2:=quotedstr(help1) ;
with ADOQuery1 do
begin
close;
```

There are also some extra buttons in this window,
- The "Unlist" button is used to unlist a medication that has been mistakenly clicked or cancelled by a customer.
- The "Record" button adds the drugs that have been sold to the electronic report.
- The "Exist" button closes this window.

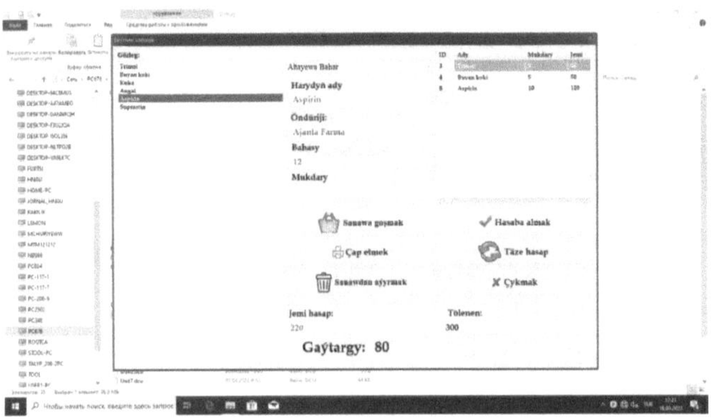

Figure 2. The interface of admin window

There is an admin window to perform basic settings in this program. Here, the administrator can enter the list of medicines, the price of medicines, and it also has a possibility to make daily, monthly and annual reports by electronical way. "Control Panel" button to be clicked on, then login and password to be entered to access it.

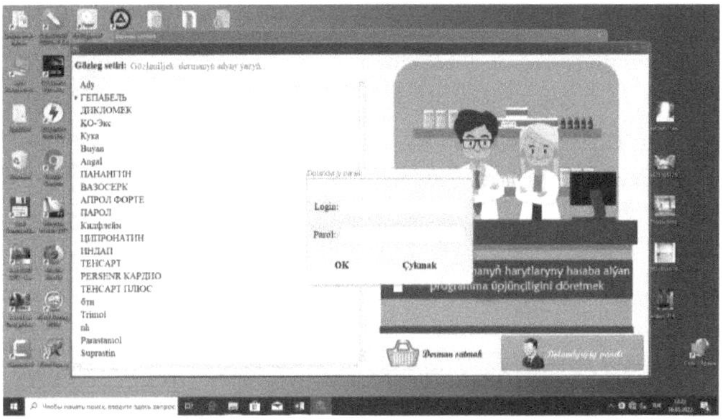

Figure 3. The interface of manager window

As you can see there are several buttons in the pop-up window:

- Registration of new product
- The report of pharmacy
- Adding a manufacturer
- Adding a new manager
- Adding a characterization to the drug
- Warehouse of pharmacy
- Adding a staff

When the first button is clicked, the following window will appear, in this window, a few lines must be filled in to register a new drug in the system, and finally, for saving must be clicked the button "Add", if the drug is to be denied, then must be clicked the button "Exit".

```
procedure TForm3.ListBox1Click(Sender: TObject);
begin
ADOQuery1.RecNo:=ListBox1.ItemIndex+1;
Edit2.Text:=ADOQuery1.FieldByName('Ady').AsString;
edit1.Text:=adoquery1.FieldByName('Öndüriji').AsString;
//Edit3.Text:=ADOQuery1.FieldByName('Type').AsString;
Edit4.Text:=ADOQuery1.FieldByName('Bahasy').AsString;
//Edit9.Text:=ADOQuery1.FieldByName('olcegi').AsString;
Edit5.CanFocus;
end;
end.
```

Medicines which have been sold for the day are showed in the Pharmacy Report window. It is possible to view daily, weekly, monthly, 6-monthly and general reports if necessary. It is also possible to print the report by selecting the required row for it. The information about the company that produces the drug is added in the “Add Manufacturer window”, it’s also has a possibility to edit information if its necessary.

```
begin
form10.adoquery1.RecNo:=i;
 ComboBox1.Items.Add(form10.adoquery1.FieldByName('isgar').AsString);
StringGrid1.Cells[0,0]:='ID';
StringGrid1.Cells[1,0]:='Ady';
StringGrid1.Cells[2,0]:='Mukdary';
StringGrid1.Cells[3,0]:='Jemi';
end;
```

Extra users to be added in this program by using “The Add New Manager” window.

“The add character to a medication” window is intended for adding information to the medication list that appears on the left side of the main window. After filling in a few lines of this code, click the Add to drug character button to save it.

A general list of drugs available in the warehouse is appeared when the Pharmacy Warehouse window opens.

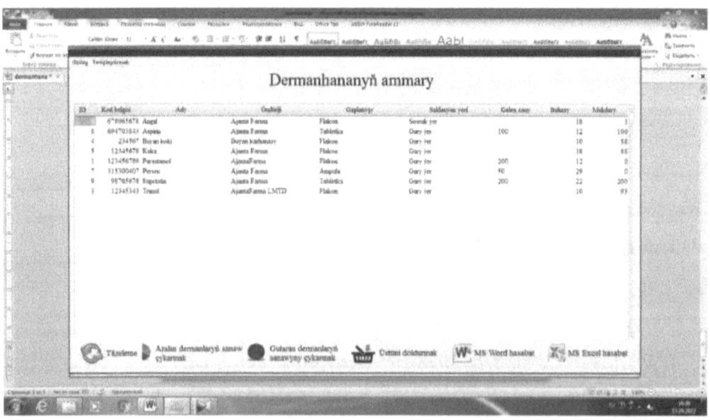

Figure 4. Pharmacy store window

In the window shown above, there is an opportunity to perform several operations without viewing the general list,

- Update, update of information;
- View the list of reduced medications;
- Show the list of expired medications;
- To fill it up;
- A possibility to report in MS Word and MS Excel.

Also in the last button- when you click on the button “Add a seller” it adds the name and surname of workers who are working as a seller in the pharmacy also it can delete and save.

References

1. Аверкин А.Н., Гаазе-Рапопорт М.Г., Поспелов Д.А. Толковый словарь по искусственному интеллекту. – М.: Радио и связь, 1992, 256 с.
2. Стюарт Рассел, Питер Норвиг. Искусственный интеллект: современный подход (AIMA), 2-е издание.: Пер. с англ. – М. : Издательский дом «Вильямс», 2006, 1424 с.
3. Умный щит. Щиты распределения электроэнергии с цифровой поддержкой. //Schneider Electric, 2015.
4. Чуриев М.М. Некоторые проблемы защиты компьютерных программ от отладчиков-дебаггеров // Аспирант и соискатель. ISSN 1608-9014, Издательство «Спутник +», Москва – 2013. – №1. – с. 64-68.
5. Чуриев М.М., Комольцев И.А. Реестр операционной системы как звено в цепочке атаки на компьютерную информацию и как средство ее защиты // Аспирант и соискатель. ISSN 1608-9014, Издательство «Спутник +», Москва – 2013. – №1. – с.76-81.
6. Стюарт Рассел, Питер Норвиг. Искусственный интеллект: современный подход (AIMA), 2-е издание.: Пер. с англ. – М. : Издательский дом «Вильямс», 2006, 1424 с.

UDC 004

IN INFORMATION COMMUNICATION TECHNOLOGIES FEATURES OF THE "ELECTRONIC MANAGEMENT SYSTEM OF DOCUMENTS" PROGRAM

M.A. Atayev, P.S. Myradov, J.R. Abdyyev
The State Energy Institute of Turkmenistan, Mary, Turkmenistan

Abstract. As of now, the education system and digital technologies have steadily established themselves on par with other industries in our country. In our State Energy Institute of Turkmenistan, in order to sustainably justify digital technologies, the center of information technologies works. In this center to automate the document circulation, the program "Electronic Document Management System" was developed. This program consists of facilitation in quick search of necessary documents, in receiving-transferring electronic documents and communication by means of e-mail among its users.

The introduction of digital technologies in the document maangement system in enterprises has great prospects. Currently, types of electronic document management are distinguished, such as production, management, archival work, organizational, secret and confidential work production, technical and technological document circulation.

There are several advantages to using electronic document management:

1. **Time saving** – electronic document management enables faster document creation, processing and searching.
2. **Increased level of secrecy** - loss of an important document or its falling into the hands of a competitor can cause significant financial damage, and electronic document management eliminates these consequences. In electronic document management, any employee of the organization is authorized to view, work on, create, and send only certain types of documents, depending on the position and level assigned to him. Any electronic document work is entered into the computer system, so any employee list can be viewed.
3. **Transparency of work production** - the system of electronic document management allows to control the creation, processing and use of documents in real time.
4. **Increased executive workload** – in paper-based workflows, employees' executive performance is low, and it takes a long time to accurately and timely execute instructions from superiors. In electronic document management, a

general manager can monitor all work processes online, which allows for increased employee productivity.

5. **Saving tools -** a large amount of money in the circulation of paper documents, its political, postal service, etc. is consumed. Electronic document management reduces the number of tools used in a business because documents are in digital form. [5]

Currently, several types of software are operating in the system of electronic document management in Turkmenistan. At the same time, it is necessary to create such programs and practical programs by our national manufacturers, national private enterprises, and national scientists and inventors from the point of view of creating a national complex in this field. The creation of such programs generally gives positive results in all areas of our country and provides the benefits we mentioned above. Presently, along with other sectors of our country, digital technologies are being actively implemented in the field of electric energy. In the Center of Information Technologies of the State Energy Institute of Turkmenistan, such an application program, i.e. an electronic document management program, was developed and prepared. There are great opportunities to use this program in all sectors of the country's industries, including the energy sector.

The program's capabilities include:

1. Establishment of electronic document circulation and mail communication among employees of the institutions;
2. To facilitate the quick finding of the necessary document;
3. Ensuring secure protection of documents;

The program is managed by an administrator, and according to the company's departments and leadership components, employee information is placed in the program's database.

The program opens as shown in Figure 1:

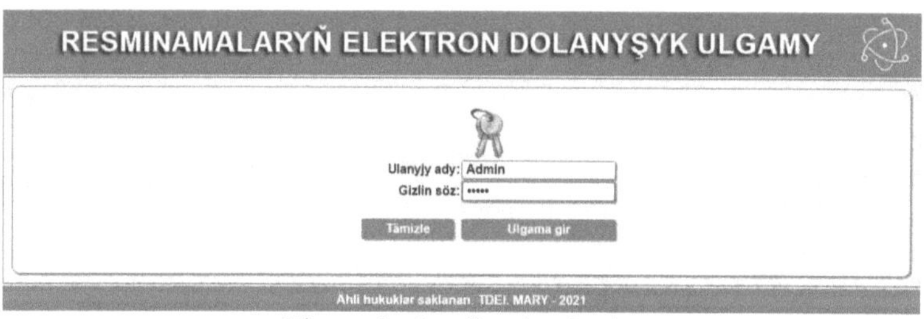

Figure 1. Application login page

Logging in to the system is done by entering a *username* and *password* as shown in Figure 1. Each user is assigned a *username* and *password* by the program administrator. This ensures secure application protection. Because it is not possible for other users to access the program. Only authorized users can use this program.

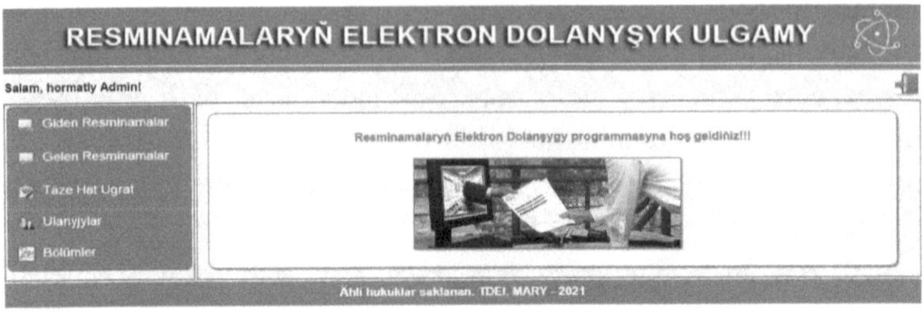

Figure 2. The home page of the program's administrator side

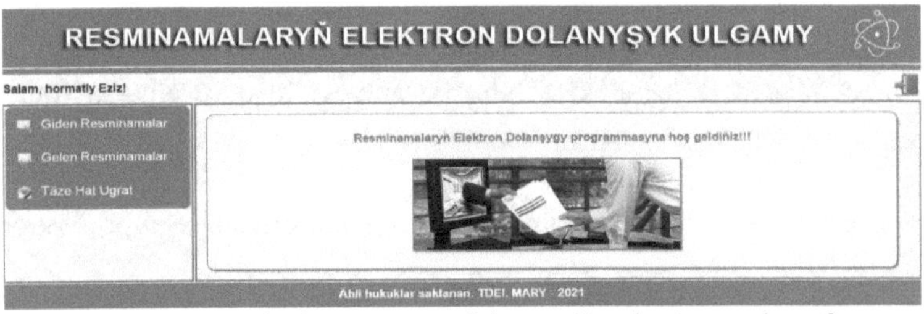

Figure 3. The home page of the application's user interface

The user rights of the program are administrator and user parties. That is, here the administrator has all rights and controls all users and departments. Therefore, it is possible to perform operations such as registering a new user or department, changing the information of any user or department, or removing it from the system. Other users are allowed to perform limited operations as shown in Figure 3. They can send new documents and view incoming documents.

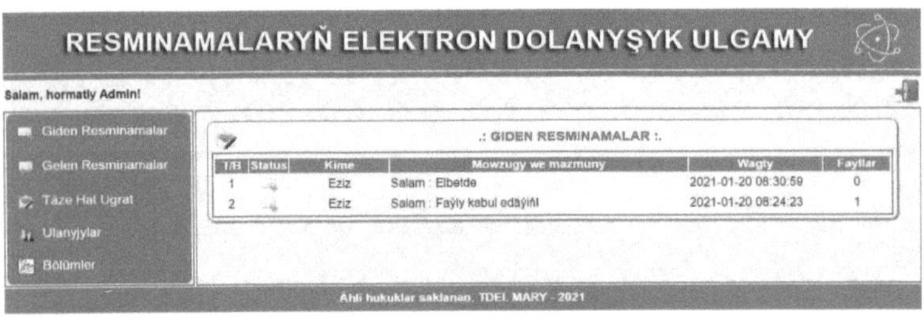

Figure 4. Department of outgoing (sent) documents by the administrator

On this page you can see the list of emails sent to the user. It shows who the message was sent to, its subject and content, the time it was sent, and a list of attached files.

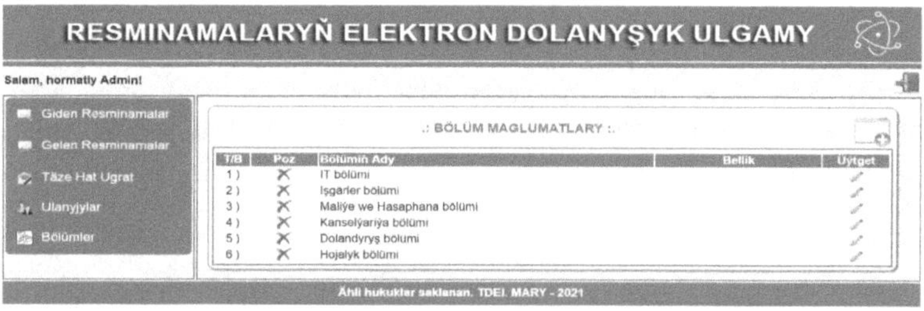

Figure 5. Sections page of the admin side

The sections listed on this page refer to sections of the company. For example, departments, personnel department, finance and accounting department, etc.

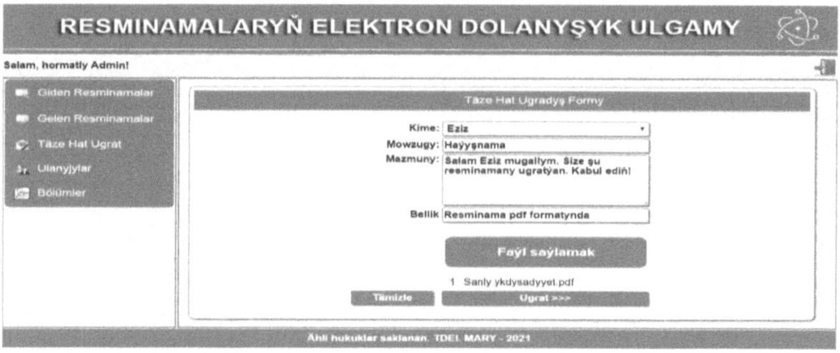

Figure 6. New document submission page by admin

On this page, the full description of the mail to be sent, the file to be forwarded is selected and forwarded to the respective user. Thus, you must first select a user in the system. Then the user should select the subject of the message to be addressed as well as the file to be sent and press the send button. A file of our choice is sent to the user along with the message we wrote.

In general, the introduction and use of the "Electronic document circulation system" program for the purpose of automating the circulation of documents in energy enterprises are among the activities carried out in the successful implementation of the "Concept for the development of the digital economy in Turkmenistan in 2019-2025", adopted by the Esteemed President, and more positively working conditions for employees will help to create and quickly implement the works.

Conclusion:

1. Creation of electronic document management software in our country;

2. Implementation of prodution into the "Electronic Document Management System" program;
3. Effective use of information and communication technologies in the digital environment.

References

1. Türkmenistanda 2019-2025-nji ýyllarda sanly ykdysadyýeti ösdürmegiň Konsepsiýasy. - Aşgabat, 2018ý.

2. Türkmenistanyň Prezidentiniň ýurdumyzy 2019-2025-nji ýyllarda durmuş-ykdysady taýdan ösdürmegiň Maksatnamasy. - Aşgabat, 2019ý.

3. Паршин К.А, Басыро Р.Р. Нормативно-правовое регулирование Российской федерации в области систем элекронного документооборота. Вестник Урфо. Безопаснотсъ в информационной сфере № 2(16)/2015г.

4. J. Abdyýew, S. Nazarow, R. Agaýew, A. Agaýew Sanly ykdysadyýet. A.: Ylym. 2021ý.

5. Умный щит. Щиты распределения электроэнергии с цифровой поддержкой. //Schneider Electric, 2015.

6. M Atayev, M Churiyev, G Atayev, A. Agayev and J. Abdiyev. Development of technology for creating software in the electric power industry of Turkmenistan. 27, March, 2023. DOI:10.1088/1755-1315/1154/1/012063.

UDC 004.4

TO CREATE AN ELECTRONIC PROGRAM "COMMUNICATION NETWORK"

M.A. Atayev, P.S. Myradov, J.R. Abdyyev
The State Energy Institute of Turkmenistan, Mary, Turkmenistan

Abstract. It should be noted that the created program arouses high interests among the pupils and students, because this program makes it possible to exchange messages through a local network. This program can be used in any factory, school, institution, office centers with a local network. The interface of the program is simple and clear in the Turkmen language, it makes easier to use and more interesting for users.

This software package is consists of two program: for the Server of Chat program and for the Client of Chat program. Let's look at first our application for the Server of Chat program. For this, we should open a new form at first and place our necessary components in it. As shown in the image above, here we need ADOConnection, ADOTable, ADOQuery, DataSource and PopupMenu components. The status of connecting server is shown in figure below.

Connecting to the server, the execution of the chatting and its program code

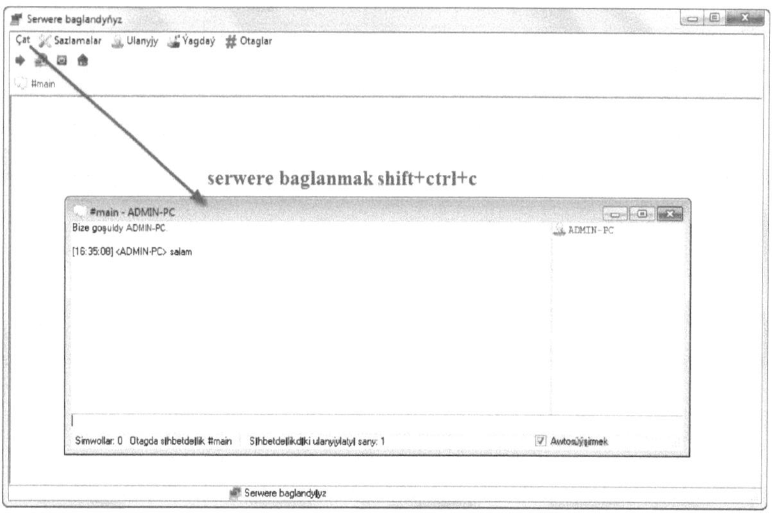

Here you have to shift+ctrl+c to connect to server or go to the settings section. In the settings section, you need to type the IP address of the user to connect to localhost. To connect to multiple computers, you need to specify the server's IP address. DBGrid, CheckBox, GroupBox components were used to display chat application settings.

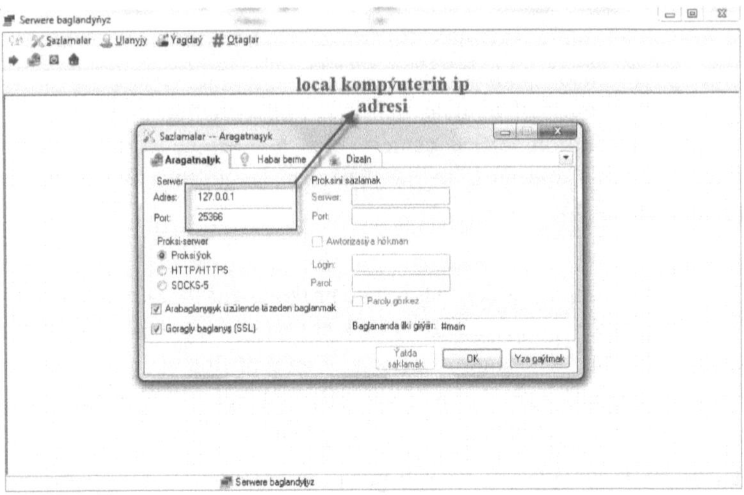


```
procedure MyDelay(ms: Integer = 1);
begin
  sleep(ms);
end;

function CheckRoom(ARoomName: string): string;
begin
  result := '';
  if trim(ARoomName) = '' then
    exit;
  if ARoomName[1] <> RoomPrefix then
    ARoomName := RoomPrefix + ARoomName;
  result := ARoomName;
end;
```

If the user wants to enter another room, he must go to the "Show Events" section in the Settings section.

The user section of the program also consists of several sections, that is, there are sections such as changing the Nickname, registering, and changing the password.

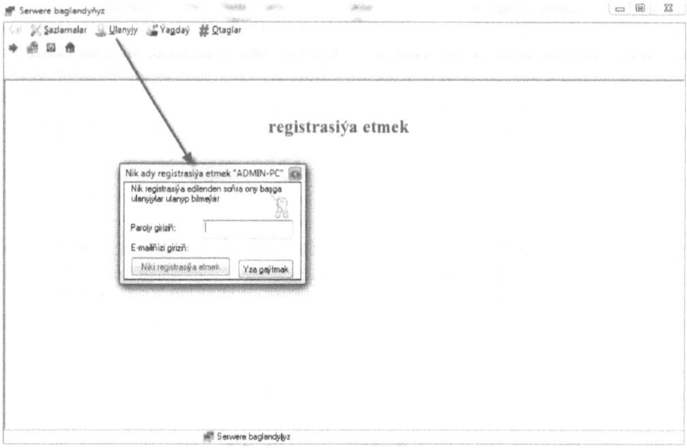


```
function IsCMDChar(achar: char): Boolean;
begin
  result := achar in ['\', '/'];
end;

function ParseUserCMD(acmd: string; FromRoom: string = ''): Boolean;
var
  cp: TUserCMDParser;
begin
  cp := TUserCMDParser.Create();
  cp.Form := MainForm;
  cp.ParseCMD(acmd, FromRoom);
  cp.Free;
  result := true;
end;
```

Information about the user is placed in the “Status” section of the program, that is, the user can write about his place of residence, e-mail, and his interests:

This program is designed to work on local computer networks.

This software package consists of two main parts, they are:

Customers

Servers

The program supply is developed as a client-server application. The main interface of this program is shown in the figure below.

Shift+Ctrl+C to connect to the server in the prepared program or go to the settings entry in the settings section. In the settings section, you need to type the IP address of the user to connect to localhost. To connect to multiple computers, you must specify the server's IP address. DBGrid, CheckBox, GroupBox components were used to displays chat application settings.

If the user wants to enter another room, then he must go to the Show Events section in the Settings section.

The user section of the program also consists of several sections, that is, there are sections such as changing the password, registering, and changing the password.

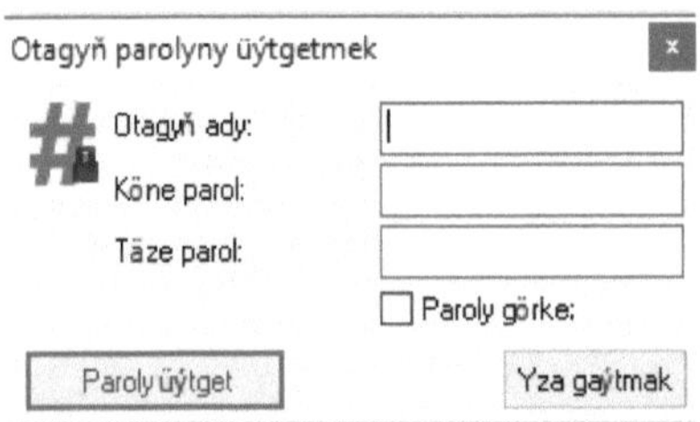

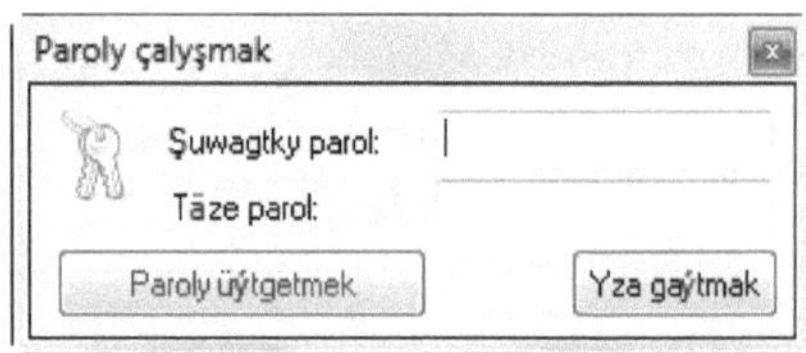

In conclusion, thus saying that all the problems considered in the project have been solved, it should be noted that the created program arouses high interest among the pupils and students, because this program makes it possible to exchange messages through a local network. This program can be used in any factory, school, institution, office centers with a local network. The interface of the program is simple, clear and in the Turkmen language, it makes easier to use and more interesting for users.

References

1. Умный щит. Щиты распределения электроэнергии с цифровой поддержкой. //Schneider Electric, 2015.

2. Чуриев М.М. Вопросы разработки современных методов защиты от троянских программ с использованием системного реестра // Сборник трудов международной научно-практической конференции студентов, магистрантов и аспирантов «рынок и эффективность производства-13». – Көкшетау, Казахстан: 2016.– с. 535-537.

3. Стюарт Рассел, Питер Норвиг. Искусственный интеллект: современный подход (AIMA), 2-е издание.: Пер. с англ. – М. : Издательский дом «Вильямс», 2006, 1424 с.

UDC 336.7

SECURITY OF CREATING A PAYMENT SYSTEM IN COMMUNICATION SERVICES

M.A. Atayev, P.S. Myradov, J.R. Abdyyev
The State Energy Institute of Turkmenistan, Mary, Turkmenistan

Abstract. The name of this software is **"Creating a Payment System for Communication Services"**. Delphi programming language and Delphi program were chosen to implement this program. This program envisages the creation of an application that automates payment services. In other words, it is planned to divide Telephone, IP-TV and Internet services according to tariffs and apply their payments in the automation program. I hope that this program will be convenient for any user and will make it easier to manage the payment system of communication services and to be informed about the types of services.

The rapid development of information and communication technologies and the impact of innovative technologies on all aspects of human activity. created the conditions for the introduction and opening of these new specialties such as: "Information-communication and innovative technologies", "Modeling and graphics", "Multimedia technologies" courses in secondary schools, "Intellectual Systems", "Expert Systems" courses in higher education schools

Today, a large number of computer programs have been created in the world. They serve a variety of purposes: graphics, engineering, video and audio processing, text editors, drivers, tutorials, tests, multimedia applications, games, web browsers, and more.

Most of the programs mentioned above are used in our country in English and Russian languages. The transfer of those programs to the Turkmen language, at least the creation of programs that make it possible to fully and comprehensibly learn how to work in those programs, will allow to attract the rest of our population to the problem of computer technology.

Since this application stores information about the user's identity and payment system, which is one of his personal data, the use of the application requires first registration and password protection. And this involves solving one of the modern global problems - cyber security.

Analytical section

In order to solve the above problem, to manage technical means and set off communication payment services, as a result, it is necessary to create a set of applications. This software suite is built in the popular object-oriented programming language Delphi.

Registration section in the program

When the program is first launched, it asks for registration. This ensures that each user's private room information remains confidentiality. Therefore, when this program is launched for the first time, the following window opens:

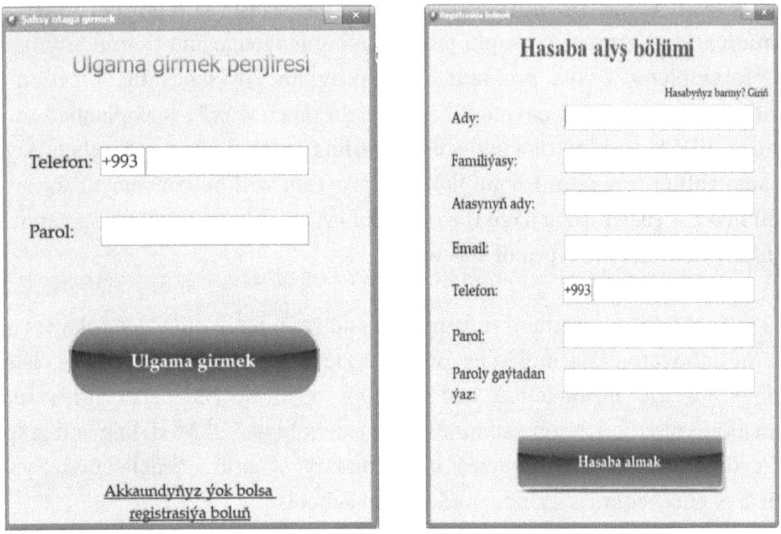

Figure 1. The program's Login and Registration windows

This site requires registration to log in. After entering the details of previous registration, you will be given the opportunity to access the home page. Here it is enough to enter the information below:

Phone number (+9936*-**-**-**)

Password protection (0000)

If you have not registered before, select the section "Register if you do not have an account" and enter the following information:

Name

Surname

Father's name

Email

Phone number

Password

Retype password

After this information is entered and saved, the user is registered in the program and an opportunity to enter the main page of the program is created.

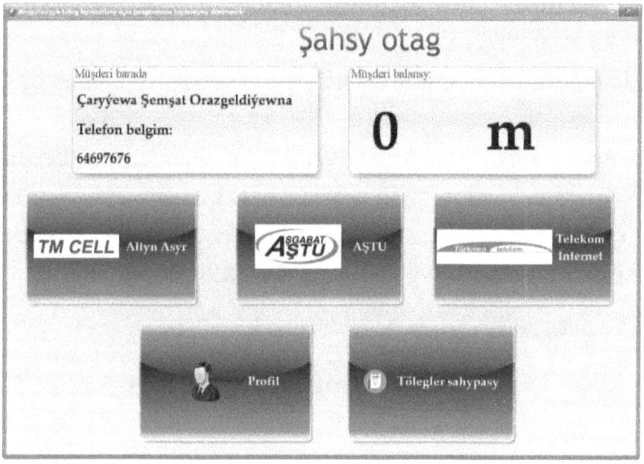

Figure 2. The program's "Private Room" windows

To prepare this form mainly TBitBtn, TDBText and TLabel, AdvSmoothButton, DBText components were used. As you can see from the picture, this page has 7 sections. They are:

About the customer
Customer balance
TM CELL "Altyn Asyr"
Ashgabat "AŞTU"
Turkmen Telecom "Telecom Internet"
Profile
Payments page

Payment window

When the system to be paid is selected, a window opens to enter the phone number and the amount to be paid. Edit, BitBtn, Label Components are used in this form. Data entered in this section is recorded in the MS Access database.

After completing the above sequence, a window will open for entering card information. After entering the information, click on the payment button, and you will be notified that the payment has been made successfully. The program code of control window for entering card information is:

In the "Altyn Asyr" (TMCELL) section of the program in our country, information about the services of this system, Internet packages, plans, USSD codes can be consult. The manuals section is displayed.

"Internet" section - in this section folders for Internet services is placed. By selecting these folders, the selected folder can be linked.

USSD codes section - in this section, USSD codes for connecting to services are placed as information.

AŞTU section - When the AŞTU section is selected on the main page of the program, the following window will open. ASHTU services available within the territory of our country are placed here. To connect to these services as well, it is possible to perform an operation by selecting the required service.

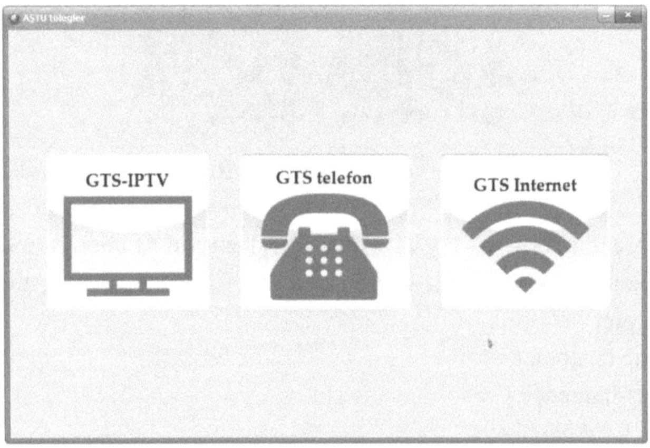

Figure 3. "Services" windows of the program

Telecom Internet Section - on this page, in our country the procedure for transferring the Telecom Internet system fee is carried out.

Profile section - to view profile information, select the Profile section. Here you can see information about the current user. If you need to change this information, make the appropriate changes and click the Save button. If you need to change the password here, it can be done using the Change password button.

Password change section - here you have to enter the existing password, enter new passwords and confirm.

Payments Page Section - to view information about payments, visit the Payments Page section. Here it is possible to see the payments of the selected day, the payments of the selected month and the payments of the selected year through a filter.

I hope that this program will be convenient for any user and will make it easier to manage the payment system of communication services and to be informed about the types of services.

References

1. Türkmenistanda 2019-2025-nji ýyllarda sanly ykdysadyýeti ösdürmegiň Konsepsiýasy. - Aşgabat, 2018ý.
2. Türkmenistanyň Prezidentiniň ýurdumyzy 2019-2025-nji ýyllarda durmuş-ykdysady taýdan ösdürmegiň Maksatnamasy. - Aşgabat, 2019ý.
3. Паршин К.А, Басыро Р.Р. Нормативно-правовое регулирование Российской федерации в области систем элекронного документооборота. Вестник Урфо. Безопаснотсъ в информационной сфере № 2(16)/2015г.
4. J. Abdyýew, S. Nazarow, R. Agaýew, A. Agaýew Sanly ykdysadyýet. A.: Ylym. 2021ý.
5. Умный щит. Щиты распределения электроэнергии с цифровой поддержкой. //Schneider Electric, 2015.
6. M Atayev, M Churiyev, G Atayev, A. Agayev and J. Abdiyev. Development of technology for creating software in the electric power industry of Turkmenistan. 27, March, 2023. DOI:10.1088/1755-1315/1154/1/012063.

UDC 004.4

CREATING SOFTWARE THAT LEVERAGES INTERNAL ASSETS

M.A. Atayev, P.S. Myradov, J.R. Abdyyev
The State Energy Institute of Turkmenistan, Mary, Turkmenistan

Abstract. The main purpose of the software is **"To create software for the sale of assets within the organization"** - to carry out the sale of tangible assets electronically in any office or enterprise. This software can be implemented in any office or business. The interface of the program is prepared in the Turkmen language. It is planned to establish electronic accounts with the help of the program. Although all information in this program is in electronic form, it is possible to change and supplement it in any time. This will help save a lot of time for the user of this program.

This software is "Creating software that transfers assets within an organization". This program is written in the Delphi programming language. The main purpose of the work is to create a program that transfers the existing property values in the office or any business. The home page of the program is shown in the image below.

As you can see from picture, there are 6 buttons on the left side of the main window:

Filling, changing, deleting the list of material assets of the organization

- Conduct property inspections
- Transferring property
- List of transferred properties
- List of properties

- About the author

The following window will appear when you click on the button " Fill, change, delete the list of material values of the institution ". The control panel of this window has several sections:

- Add
- Change
- Disconnect
- Search string
- To export

The program code for filling the list of material assets of the institution, changing and removing the window:

```
procedure TForm6.MSWord1Click(Sender: TObject);
var
word1:Twordapplication;
   doc:variant;
   template:olevariant;
   i:integer;
begin
   try
  word1:=twordapplication.create(self);
  template:=ExtractFilePath(Application.ExeName)+'nusga.docx';
  word1.Documents.Add(template,emptyparam,emptyparam,emptyparam);
  doc:=word1.ActiveDocument;
  word1.Visible:=true;
except
  showmessage('word doredilmedi');
end;
word1.Selection.Start:=0;
with ADOQuery1 do
  begin
for i:= 1 to RecordCount do
   begin
RecNo:=i;
doc.Tables.Item(1).rows.add;
doc.Tables.Item(1).cell(i+3,1).Range.Text:=IntToStr(i);
```

```
    doc.Tables.Item(1).cell(i+3,2).Range.Text:=FieldByName('Tabel_nomeri').asstring;
    doc.Tables.Item(1).cell(i+3,3).Range.Text:=FieldByName('Ady').asstring;
    doc.Tables.Item(1).cell(i+3,4).Range.Text:=FieldByName('Sany').asstring;
    doc.Tables.Item(1).cell(i+3,5).Range.Text:=FieldByName('Bahasy').asstrin;
    // Word tekst redaktorynda maglumat gorundaky ýazgylary geçirmek üçin ýazylan kod.
      end;
    end;
```

When you click on the "Add button", the following window will appear. In this window, the addition of material values is performed. When entering material values, there is also an option to add an image. After entering all information, click on the "Save" button.

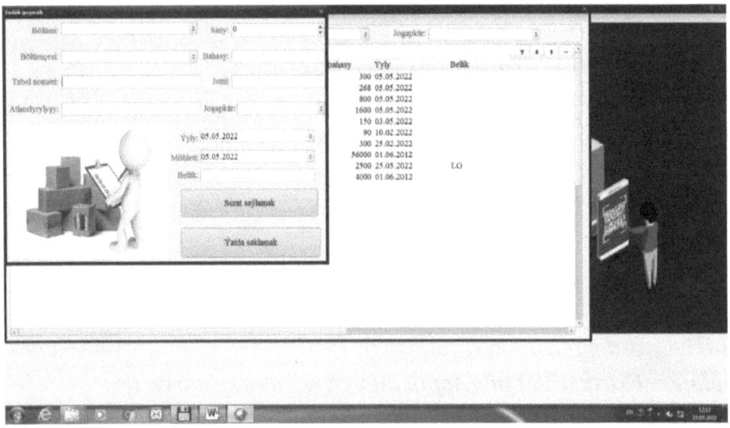

By clicking on the "Change" panel, the relevant location, name, name of the responsible person, price, serial number, etc. of the existing material values will be displayed. Is given the option to change. After making the changes, click on the "Save" button.

When you click on the "Remove" button, it is possible to delete the existing material values in the list. For that, you have to select the required purchase and click on the "Remove" button.

When this Search Bar panel is selected, the action to show or hide the search bar is performed. If the search bar is active, a search bar will appear at the top of the window.

When the "Export" tab is selected, it is possible to export the existing list as a file to an MS Word or MS Excel editor. In the image below, the list of material values has been exported as a table to an MS Word document.

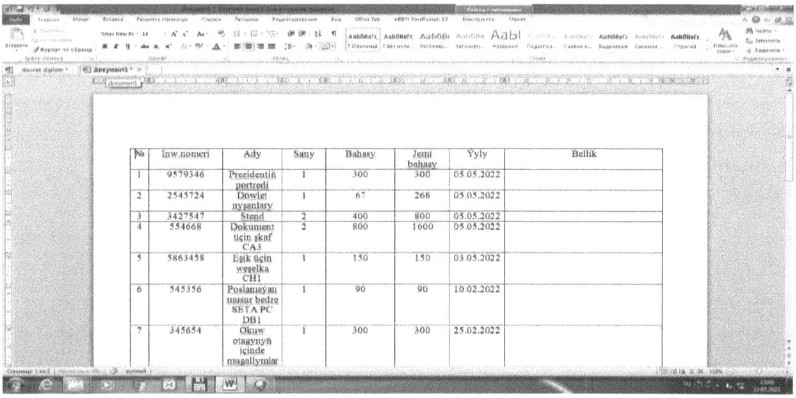

1. When you click on the button “Perform property inspection”, the following window will appear. This window will display all registered material values. Optionally, the material values of the required section can be viewed separately in this window. There are also lines are used to differentiate existing, out of line, and broken property values.

2. When you click on “Transfer property” button, the following window will appear. This window can be used when the user needs to transfer material values to another location. To perform the operation, you need to select the necessary monetary value and fill in the lines. After the process is done, you need to click on “Checkout” button.

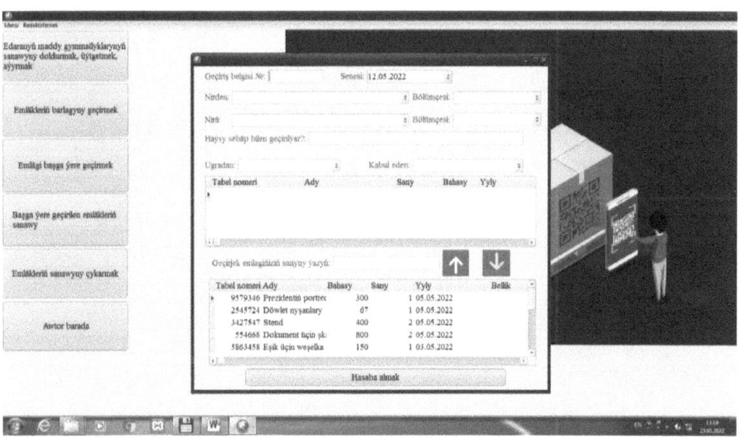

3. “List of transferred Assets”. This section shows the result of the above operation.

4. The image above shows the window that appears when you click the "Extract Property List" button. This window contains several buttons:

- Remove grouping by section
- Grouping by department
- Calculating the total value of material values
- Calculating the value of selected tangible assets
- View lists

"Clicking on the "View Lists" button will bring up the following window, which will display a list of material values and provide general information about them. There are several buttons in this window, that is, with the help of them, it is possible to export the existing list as a file to MS Word or MS Excel editor, and there is an option to extract the list of expired properties.

References

1. Marz N., Warren J. Big Data: Principles and Best Practices of Scalable Real-Time Data Systems. Manning Publications, 2015.
2. Knaflic, C. N. (2015). Storytelling with Data: A Data Visualization Guide for Business Professionals. Wiley.
3. Tufte, E. R. (2001). The Visual Display of Quantitative Information. Graphics Press.
4. Nakamoto, S. (2008). Bitcoin: A Peer-to-Peer Electronic Cash System.

UDC 004.056

CYBER SECURITY TERMS DICTIONARY

M.A. Atayev, P.S. Myradov, J.R. Abdyyev
The State Energy Institute of Turkmenistan, Mary, Turkmenistan

Abstract. Title of my graduation project is " Cyber Securyty Terms Dictionary." This project involves the creation of software for the Turkmen-English-Russian dictionary of terms on Cybersecurity. The program should not only display the translation of words from one language to the other two, but also show the interpretation of this term in three languages.
The program should be able to edit terms and their interpretation, enter new terms, download them from outside the Excel file (* .xls), attach and view the figure according to the topic of the term Cybersecurity.

The work of our program consists of the following stages:

- Manual introduction of new terms and their interpretations;
- Downloading data from Excel file;
- Sorting and filtering in different languages
- Loading pictures.

Manual introduction of new terms and their interpretations

Below are the components for viewing terms in three languages.

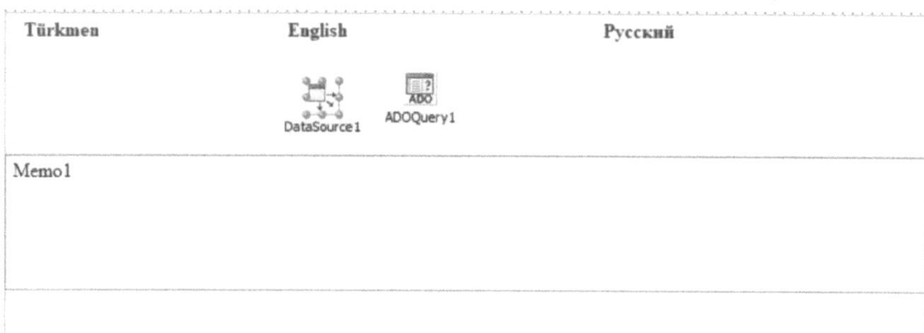

ADOQuery1: TADOQuery - to connect to AdoConnection1 (located on the main form) and connect using Sozler table SQL queries

DBGrid1: TDBGrid - for visual display of a table with classrooms, has four fields – Turkmen, English, Russian words.

DataSource1: TdataSource - for connecting DbGrid1, DBEdit1, DBEdit2, DBEdit3, DBMemo1, DBMemo2, DBMemo3, DBNavigator1 to ADOQuery1

Below are the components for entering and editing terms and their definitions in three languages

To perform these functions, the window has the following components:

DBNavigator1: TDBNavigator - to manage the table (create and delete, edit, save a record, navigate through records)

DBEdit1: TDBEdit - for entering terms in Turkmen language
DBEdit2: TDBEdit - for entering terms in English
DBEdit3: TDBEdit - for entering terms in Russian language

DBMemo1: TDBMemo - to enter the interpretation of the term in the Turkmen language
DBMemo2: TDBMemo - to enter the interpretation of the term in the English
DBMemo3: TDBMemo - to enter the interpretation of the term in the Russian language

Downloading data from Excel file

To download data from Excel files, the following components are included:

OpenDialog1: TopenDialog - to open an Excel file

StatusBar1: TstatusBar - to show the progress of the process of load

Sorting and filtering in different languages

Loading pictures

The result of this graduation project is software, the main window is shown in the figure below.

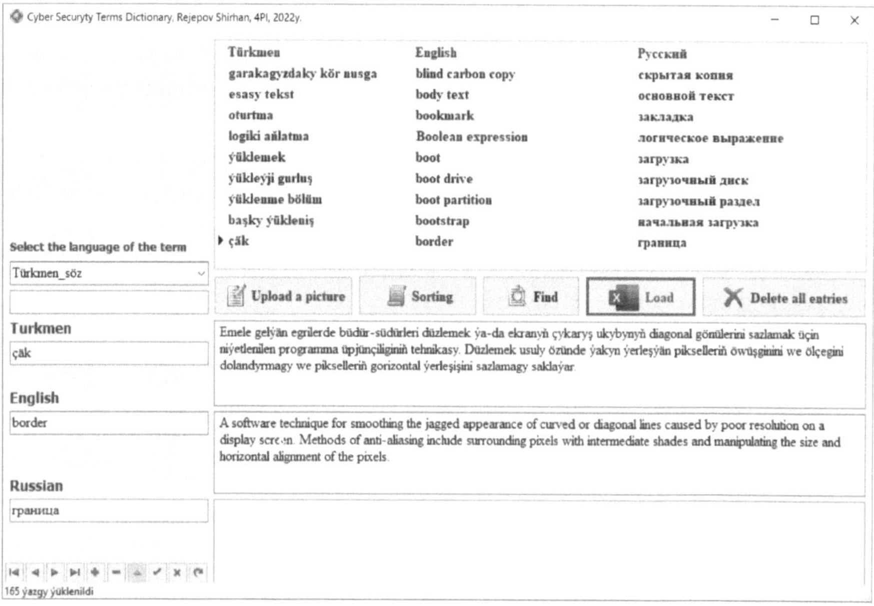

At the top of the program is a table of terms in three languages - Turkmen, English and Russian.

To filter terms you do not need to choose which language to filter in the Select the language of the term field:

When you press the Sorting button, all terms are automatically filtered, depending on the language you have chosen, alphabetically, for example in Turkmen:

In addition to selecting a sorting language, you need to specify which part of the search term or terms you want to search for. For example, in the Turkmen language we will search for the word “kiber”:

The part of the program specified in the figure is intended for manual input of terms and their interpretation.

To load terms and interpret them from an Excel file, click the Load button. The Excel file selection dialog opens.

After that, loading the terms and their interpretations into the corresponding fields will begin. The boot course will be displayed in the lower status bar.

It should be noted that the structure of the downloaded Excel file must match the figure.

Now let's look at an Excel file that will solve the terms and their interpretation in Turkmen, English and Russian.

This cell B - contains the term English, cell C - contains the term Turkmen, cell D - contains the term Russian, cell E - contains the interpretation of the English language, cell F - contains the interpretation of the term Turkmen - the content of the G, the cell Russian language.

It is possible to upload a drawing to terms that will be displayed when you select this term. To do this, click the Upload a picture button. The Picture Selection dialog box opens. You need to select an available drawing and it will load into the program.

After downloading the image, you need to save it to the database by pressing the corresponding button on the navigator.

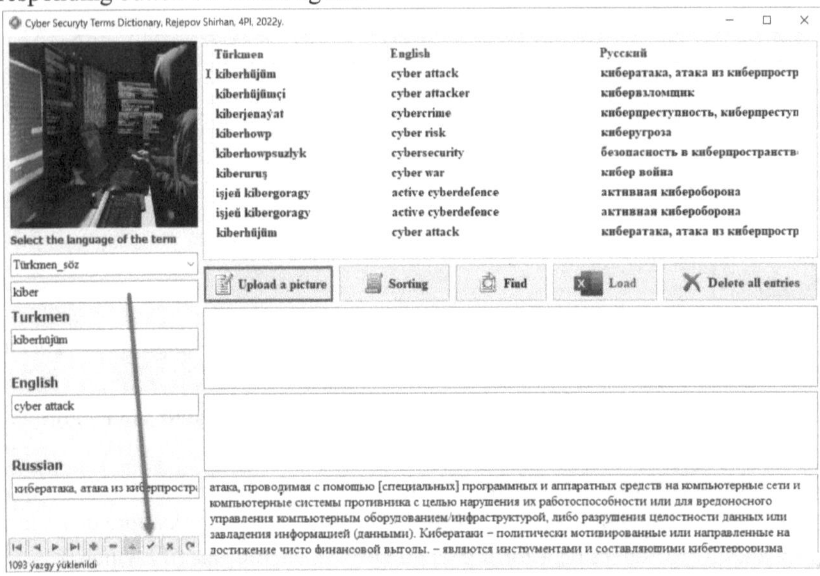

To view a larger image, you need to double-click on the image.

Click Delete all entries to delete all terms.

In this way, all the goals and objectives set out in the diploma project have been successfully fulfilled.

References

1. Глушаков С.В., Клевцов А.Л., Программирование в среде Delphi 7.0, Фолио. 2003. 415 с.

2. Попов В.В. Программирование в Delphi. Оптимальный подход, Век, 2005. 150 с.

UDC 004

"OKUWÇY" DIGITAL SOLUTION OF THE COMMODITY CIRCULATION

M.A. Atayev, P.S. Myradov, J.R. Abdyyev
The State Energy Institute of Turkmenistan, Mary, Turkmenistan

Abstract. This software asks the user for a password when first launched, which prevents any user from using the software. The password can be changed by the app owner. This is in line with cyber security requirements. It can also be installed on all computers at points of sale. Another feature of the software is that you can manage goods in the warehouse by connecting software installed on each computer, called "Okuwçy" digital solution of the commodity circulation".

This program covers various areas of commerce. In order to control and manage the security of the commodity circulation and the duration of the work performed on the computer server through the provided to the customers and how many manats and which goods are sold on the computer screen, the software called ""'Okuwçy" digital solution of the commodity circulation" has been developed. The software was developed entirely in the Delphi programming language.

Today, educational institutions, educational centers, enterprises, workplaces, shops and outlets of the country are equipped with modern computer technologies and the latest achievements of science. The computerization of trading points and the establishment of commodity circulations also require their protection. With the help of various applications, it is possible to prevent or limit the theft of data by strengthening and improving the security of the commodity circulation system. It is advisable to use the """Okuwçy" Digital Solution of the Commadity Circulation" program in order to ensure that the commadity circulation of those computers are suitable during work, to carry out such tasks as out-of-work situations, removing the sold goods from the warehouse, and transferring the money of the sold goods to the treasury.

This software is mainly designed to monitor what and how many products are available in stores, outlets, school supply stores and warehouses.

Along with the installation of computer technology in all parts of our country, it is widely used in stores, trade points, school supplies stores, warehouses and all sectors of industry.

As well where computer technologies are available, i.e. installed, software installation works are carried out on computers in stores, trade points, school supply stores and warehouses, taking into account the benefits of time saving and economic benefits. Of course, it is also necessary to manage and monitor it through the installed software. This is because a number of tasks are planned so that the accounts of the movement of goods are not lost and the accounts of the goods are kept on time.

This software consists of the following components:

- Sell;
- Income-Output Add ;
- Correction Entry;
- Cancel;
- Commodity (Goods);
- Statistics;
- Printing;

taking into account the above, the program ""Okuwçy" Digital Solution of the Commodity Circulation" can be widely used to perform, control and manage several processes through the departments placed for the purpose of digitalization of commodity circulation.

One of the main features of the developed software "Okuwçy" digital solution of the commodity circulation is the possibility to manage each product by BarCode and for the purpose of carrying out commodity circulation in the working window of the program, BarCode, Product Information, Quantity, Purchase Price, Selling Price , Benefit and Date of Writing fields are located. It contains all the information about the goods sold on this site. By the general manager of the store or trading point, it is possible to remotely monitor the flow of goods and the transactions being carried out.

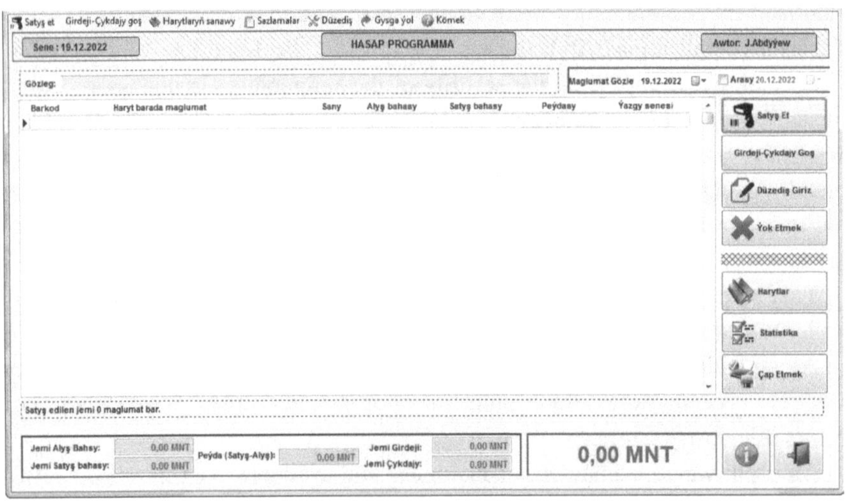

Figure 1. "Okuwçy" solution of the commodity circulation" is the main business window of the software

In the sales department, you need to enter some information about the product into the database. It contains some information such as the name of the item, the purchase price of the item, the selling price of the item, the quantity of the item in stock, and the date the item was purchased and sold.

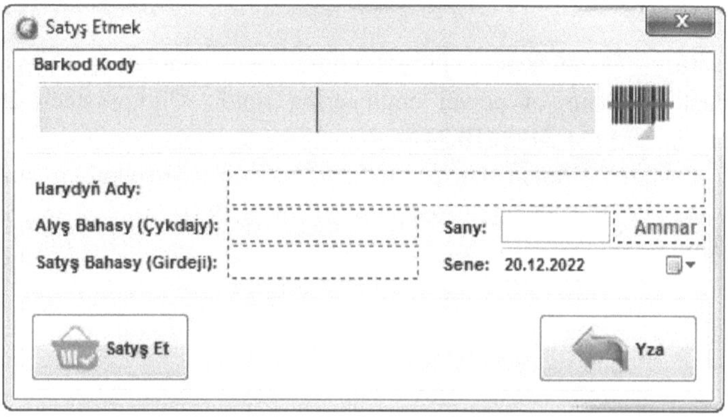

Figure 2. About the commodity in the "Sell" task window contains information

"Income-Output Add" - by clicking the button, it is possible to calculate the income and expenses of the transaction in separate manats.

"Correction Entry" - in the section, it is possible to change and correct the information of the goods in the warehouse.

"Cancel" - in the section, it is possible to cancel any purchase by the warehouse manager. If you want to cancel, you must select the purchase and click the "Cancel" button.

"Commodity" (Goods) - the section contains information about all goods stored in the warehouse. It shows the inventory order number, Barcode, Commodity name, quantity, purchase price, selling price, and net profit. After entering the expiration date of each item in the warehouse, it automatically reports until the expiration date is finished.

"Statistics" - the section contains information on trade. It calculates total cost of goods sold, total cost of goods purchase, and gross profit separately. Through this section allows you general account calculate all of them together, rather than each purchase separately. There is no specific limit on the number of goods in the commodity circulation. Through this software accurate and precise account management is possible.

References

1. Чуриев М.М. Вопросы разработки современных методов защиты от троянских программ с использованием системного реестра // Сборник трудов международной научно-практической конференции студентов, магистрантов и аспирантов «рынок и эффективность производства-13». – Көкшетау, Казахстан: 2016.– с. 535-537.

2. S. Nazarov, B. Jumayev. Programming-based laboratory assignments for undergraduate students of power engineering fields. Compsystech 2021. Ruse, Bulgaria. DOI: 10.1145/3472410.347429

3. Рассел Стюарт, Норвиг Питер, «Искусственный интеллект: современный подход (AIMA)», 2-е издание.: Пер. с англ. – М.: Издательский дом «Вильямс», 2006. – 1424 с.

UDC 82

COLLECTION OF POEMS BY MAKHTUMKULI FRAGI SOFTWARE FOR SMART DEVICES

M.A. Atayev, P.S. Myradov, J.R. Abdyyev
The State Energy Institute of Turkmenistan, Mary, Turkmenistan

Abstract. This software helps every reader to become familiar with the poems of the poet of the Turkmen people, Magtymguly Pyragy, and conveys the poet's wisdom not only to the Turkmen people, but also to people from all over the world, giving them the opportunity to study the work of this great poet.

The main focus of the President's science and education policy is to attach great importance to the new achievements of science and technology at the world level in the country. One of the main conditions for the further development of the Turkmen people in a noble and happy life is the development of science and technology and the introduction of advanced technology into the national economy. For this purpose, great efforts are being made in our country for the young generation to develop in all aspects, become experts in the world's most advanced magic and magic systems, master the achievements of science at the world level in time, and be able to keep pace with it. The same work is done in our university.

It is fully established for students to independently use modern technologies during the educational process, and to create software that supports the modern national interface. In this software, Magtymguly Pyragy's poems are compiled in four languages and all the possibilities to read them in electronic form have been created. This software will be a rich literary school not only for the Turkmen people but also for the peoples of the world.

Under the wise leadership of President, the rich and glorious historical and cultural heritage of the Turkmen people is studied on a scientific basis, and large-scale works are being carried out to preserve and transmit it to our generations. Their importance in educating our young generation in terms of patriotism, humanism, morality and aesthetics is very important. Accordingly, protection of the historical and cultural heritage in the country, study and transmission to young generations, as well as wide promotion is one of the priorities of the state policy.

Magtymguly's poetic-philosophical literary legacy is a song of great love for the Great Creator, the Motherland, man, nature, and human life, and it took an eternal place in the hearts of the people. If the value of his creativity left to us is evaluated only in terms of artistic words and poetry, if Magtymgula is considered only as a poet, then it will be neglected. He is a wise poet who developed the philosophical ideas not only of the Turkmen people, but also of the world, enlightened the artistic mind, and artistically reflected the secular life. Magtymguly united his philosophical views on the world, humanity, Motherland, love with sensitive human feelings, expertly used the most perfect ways of thinking, penetrated into the human soul and took a permanent place in the people's history.

The name and symbol of the year 2024 "Ocean of Wisdom Magtymguly Pyragy" is placed on the main page of the program. A new poem by the National Leader of the Turkmen people called "The Ocean of Wisdom Magtymguly Pyragy" is posted. Each word and line of this poem has a very deep meaning. The translation of the great poet's collection of poems into four languages provides an opportunity to get acquainted with the Turkmen culture, spirituality, and moral principles through the software.

Figure 1. Home page of the program

In the "Additional menu" of the program there are 6 sections:

1. Life and work

2. Poems

3. Your preferences

4. Language selection

5. Theme

6. Share the app

In the "Life and creativity" section of the program, some information about Magtymguly Pyragy is given. The life and creativity of the master of words Magtymguly Pyragy and his poems are presented.

Figure 2. "Life and Creativity" section of the program

In this software, the poems of our great poet have been translated into four languages. The whole world has made it easier for readers to collect the poems of the Turkmen poet Magtymguly Pyragy in electronic form and in the Turkmen, Russian, English and Turkish languages. Nowadays, due to the further development of digital technology, it is possible to create different types of software. As part of the digital system, software for the collection of poems of Magtymguly Pyragy, the great master of words, was also created. This software is now enabled to work on any android cell phones. The program will be a literary school and a literary resource for young people studying in preschools, middle schools, vocational schools and universities.

In the "Poems" section of the software, there are pictures of the poem "Turkmen" translated into four languages. There are poems in Turkmen, Russian, English and Turkish languages. In this poem, Magtymguly Pyragy is one of the deeply meaningful poems written about the Turkmen people, the brave sons of Turkmen and the Turkmen land. The poems of Magtymguly Pyragy, a true patriot of the Turkmen land and an ardent waspist, serve the spiritual perfection and moral health of our society. The thoughts of the great wordsmith in his poems, which became valuable to all mankind, and his exhortations about patriotism, humanitarianism, and moral purity, are the school of life of our people.

This software enables every reader to become familiar with the poems of the poet of the Turkmen people, Magtymguly Pyragy, and conveys the poet's wisdom not only to the Turkmen people, but also to people from all over the world, giving them the opportunity to study the work of this great poet.

References

1. Давыдов, Станислав IntelliJ IDEA. Профессиональное программирование на Java. Наиболее полное руководство (+ CD-ROM) / Станислав Давыдов , Алексей Ефимов. - М.: БХВ-Петербург, 2011.

2. M. Ataýew, B. Atajanow. Magtymguly the infinite light of the human soul. Актуальные вопросы современной науки сборник статей VI Международной научно-практической конференции, Состоявшейся 20 апреля 2023 г. в г. Пенза.

3. M. Atayev, M. Churiyev, G. Atayev, A. Agayev and J. Abdyyev. Development of technology for creating software in the electric power industry of Turkmenistan. 27, March, 2023. DOI:10.1088/1755-1315/1154/1/012063.

УДК 378.147

МЕТОДЫ ПРЕПОДАВАНИЯ В УНИВЕРСИТЕТЕ

М.А. Атаев, П.С. Мырадов, Дж.Р. Абдыев
Государственный энергетический институт Туркменистана
Мары, Туркменистан

Аннотация. Компетентностный подход обусловил необходимость обоснования и внедрения в профессиональную деятельность новых (или усовершенствования существующих) ролей преподавателя вуза. Для их выявления в статье проведен анализ научных и монографических публикаций отечественных ученых и синтез практических разработок. В результате обобщения полученных данных на компетентностных основаниях получена взаимосвязь профессиональных ролей с основными видами деятельности преподавателя вуза. Представленные в статье результаты могут быть использованы в практике повышения квалификации преподавателей высшей школы для обеспечения высокого уровня их профессиональной деятельности и разработки профессионально квалификационных характеристик, обеспечивающих возможность рейтингования преподавателей.

Ключевые слова: преподаватель вуза, компетентностный подход, профессиональные роли преподавателя.

Подготовка специалистов с высшим образованием по различным направлениям и профилям относится к одной из самых сложных категорий. Поэтому целесообразно осуществлять пристальное и постоянное внимание ко всем факторам, которые непосредственно влияют на качество высшего профессионального образования. И, конечно, в первую очередь, – это особое внимание к ключевой фигуре образовательного процесса вуза – преподавателю, который всегда играл, играет и сохранит в будущем ведущую роль в высшем образовании. Именно преподаватель осуществляет основные и стратегические функции в процессе профессионального развития личности студента и его самореализации в образовательном процессе высшего учебного заведения.

В настоящее время возрастает роль активных методов обучения в процессе подготовки будущих специалистов к межкультурному взаимодействию, так как пассивные методы обучения неэффективны в сфере выработки устойчивых знаний, умений, навыков и профессиональных компетенций, обучающихся в ходе учебного процесса. Заучивание лекционных материалов и содержания учебников не даёт практической составляющей, и таким образом, студенты приобретают теоретические знания, не умея применять их на практике в реальной жизни. В отличие от пассивных, активные методы обучения способствуют самостоятельному познанию мира обучающимися в ходе активности на занятиях и во внеурочное время, а также помогают развивать такие немаловажные для управленца качества как: способность самостоятельно принимать решения и нести за них ответственность, действовать в быстро изменяющейся обстановке, приближенной к реальной и т. д.

Таким образом, проработав с помощью активных методов обучения ряд проблемных ситуаций и конкретных путей их решения, в дальнейшем будущие специалисты учатся творчески подходить к поставленным задачам и самостоятельно искать нестандартные решения, наиболее эффективные в каждом конкретном случае. Всё это особенно актуально в связи с 100 постоянным ростом требований к менеджерам, как к специалистам, способным понимать и принимать культурное разнообразие современного мира.

Ниже приведены некоторые условия, которые могут оказать положительное влияние на качество преподавания:

1. **Заинтересовать учащихся предметами, объяснить им необходимость быть профессионалом - показатель профессионального мастерства преподавателя.** В целом понимание предметной оценки у студента более развито, а отношение к учебе формируется позже. Урок преподавателя должен мотивировать студента выбрать правильное место в жизни. Наличие высококвалифицированного, современного специалиста и его преимущество в жизни мотивирует студента к углубленному изучению своего предмета. Увлеченность студента избранной профессией и уверенность в своих силах в определенной степени зависят от преподавателя. Преподавателя должны всегда стремиться к увеличению разнообразия обучения и обогащению содержания уроков. Анализируя международный опыт совершенствования образования, учащиеся за рубежом отмечают, что методы организации занятий и предоставляемая в них информация устарели. Поэтому преподавателю следует на уроках приводить примеры, чтобы слушатели могли их понять, и рассказывать о своем опыте работы.
2. **Возможности инновационных технологий и Интернета отлично подходят для самостоятельной учебы и обучения студентов.** В частности, они делают более доступным для студентов получение необходимых знаний самостоятельно. В современных условиях преподаватель должен не только дать учащимся основы знаний, но и подготовить их к самостоятельному обучению, творческому мышлению и

принятию правильных решений. Овладение знаниями должно стать собственным трудом студента. Сегодняшние достижения науки и технологий, методологии повышают самостоятельную подготовку студентов. Приобретение привычки к самообучению и самообучению – одно из ключевых условий достижения совершенства. Преподаватель должен приложить все усилия для самостоятельного развития и совершенствования профессиональных навыков учащихся с помощью новых методов и информационных технологий.

3. **Задача образовательных учреждений – повысить мотивацию учащихся к участию в соревнованиях.** Конкуренция мотивирует людей работать усерднее обычного, ответственно подходить к своей работе. Для создания такой ситуации среди студентов целесообразно организовывать конкурсы талантов. Соответствующим руководителям вуза следует сформировать среди студентов смешанные группы, назначить к ним наставников, присвоить каждой группе особые названия, а затем регулярно организовывать конкурсы талантов среди этих групп в определенный день недели (раз в неделю или один раз в две недели). Выбор тем конкурса на основе предложений, учащихся в цифровой системе, предоставление определенного времени на подготовку, предложение ссылок на специальные книги, научные труды, статьи по выбранной теме повышает активность молодежи. Подготовка мотивационных материалов и видеороликов о каждой группе повышает интерес учащихся. Результаты мозгового штурма, очки, заработанные каждой командой, и другие показатели должны фиксироваться на протяжении каждого сезона.
4. **Метод опроса в контексте высшего образования. Анкеты используются для различных целей.** Часто в международном опыте этот метод используется для улучшения учебы студентов в рамках высшего образования, изучения их взглядов и предложений. Главное условие эффективности опроса общественного мнения – не раскрывать личности тех, кто его заполняет. Опрос эффективен, если он проводится с использованием современных технологий и возможностей цифровой системы образования. Студент имеет возможность полно и открыто высказывать свои ответы и предложения на вопросы, заданные в электронном виде в цифровой системе. На основании полученных ответов и предложений проводится анализ, полученные результаты используются при планировании внутренних дел, принятии решений, совершенствовании организационной деятельности. Принцип поощрения педагогических навыков учащихся должен достигаться путем

налаживания с ними дружеских отношений, а также привлечения к педагогической и методической деятельности опытного, состоявшегося студента. Взаимодействие учащихся является более интимным, чем взаимодействие преподавателя и студента. Эта ситуация позволяет студенту с низкой успеваемостью открыто и без колебаний обсуждать свои трудные вопросы со своим партнером.

5. **Основная задача высших учебных заведений – воспитать современную туркменскую молодежь, дать ей образование на международном уровне.** В настоящее время наше государство создает широкие возможности в этой сфере. Устанавливает позитивные связи с высшими учебными заведениями и научными учреждениями зарубежных стран. Молодежь отправляют учиться за границу. Сотрудничество нашей молодежи с зарубежными специалистами в области образования и науки в центрах, где реализуются достижения передовой науки и техники, играет важную роль в формировании национальных специалистов международного уровня.
6. **Дополнительные занятия.** Дополнительные уроки позволяют преподавателю показывать внеклассные видеоматериалы, официальные стенограммы и полные интервью. На дополнительных уроках темы освещаются по желанию учащихся более углубленно, эффективно (видео записи (презентации, выступления, конкурсы) можно преподавать.
7. **Регулярная организация ознакомительных посещений музеев, театров, выставок, природных достопримечательностей направлена на повышение знаний учащихся и расширение их мировоззрения.** В отличие от обычных занятий, организация любознательных занятий и поездок за пределы образовательного центра обеспечивает более эффективный и интересный прием. новой темы. Наряду с обучением путешествия расширяют кругозор учащихся. Экскурсии дают возможность привлечь молодежь и дают возможность учителям оценить, как учащиеся справляются в новой среде.
8. **Использование современных информационно-коммуникационных и компьютерных технологий в образовании и обучении способствует повышению качества образования.** Внедрение цифрового образования в образовательных учреждениях и использование технологий дистанционного образования обеспечивают новые этапы развития. Цифровое образование – потребность времени. Преподавателя должны всегда быть в курсе цифровых технологий и уметь в полной мере использовать их преимущества. Ресурсы и услуги электронного обучения стали доступны учащимся учебных заведений дома и даже в дороге.

С развитием научно-технического прогресса, увеличивается объем информации, обязательной для усвоения. Установлено, что информация быстро устаревает и нуждается в обновлении. Отсюда вытекает следующее, что обучение, которое ориентировано главным образом на запоминание и сохранение материала в памяти, уже только отчасти сможет удовлетворять современным требованиям. Значит, выступает проблема формирования таких качеств мышления, которые позволили бы студенту самостоятельно усваивать постоянно возобновляющуюся информацию, развитие таких способностей, которые, сохранившись и после завершения образования, обеспечивали человеку возможность не отставать от ускоряющегося научно-технического прогресса.

Из этого можно сказать, что нужны новые методы и подходы в обучении, которые могли научить студентов учиться, т.е. самостоятельно находить и усваивать нужную информацию. Ведь, то, что усвоено самостоятельно, методом проб и ошибок усваивается лучше. Использование преподавателями активных методов в вузовском процессе обучения способствует преодолению стереотипов в обучении, выработке новых подходов к профессиональным ситуациям, развитию творческих способностей студентов.

Список литературы

1. Союнова Г., Непесова Х., Мамметдурдыев А. МЕТОДЫ ОБУЧЕНИЯ ЭКОНОМИЧЕСКОЙ ТЕРМИНОЛОГИИ В ВУЗАХ ТУРКМЕНИСТАНА //Вестник науки. – 2023. – Т. 2. – №. 5 (62). – С. 155-157.

2. Гребенев, И.В. Формирование педагогического профессионализма преподавателей в условиях классического университета: на материале подготовки преподавателей физики: дис. ... д-ра пед. наук. – Нижний Новгород, 2008.

УДК 575.112

БИОИНФОРМАТИКА КАК ФОРМА БИОТЕХНОЛОГИИ

М.А. Атаев, П.С. Мырадов, Дж.Р. Абдыев
Государственный энергетический институт Туркменистана
Мары, Туркменистан

Аннотация. Компьютерные программы пишут на самых разных языках программирования: в машинных кодах, на ассемблерах ил же на языках выского уровня. Программы, написанные на ассемблере или языке высокого уровня, должны быть преобразованы в машинный код путем ассемблирования и компиляции. В операционной системе Windows файлы в машинном коде называются исполняемыми файлами. А соответствующие файлы в системе Unix-исполняемыми образами. Такие файлы непосредственно выполеяются процессором компьютера.

Ключевые слова: биоинформатика, язык программирования, развитие ИТ.

Введение. Под биоинформатикой обычно понимают решение задач биологии с использованием компьютеров и вычислительных технологий. Сегодня большинство этих проблем относится к области молекулярной биологии. В результате в этой области собрано огромное количество данных за более чем 20-летний период. Примером может служить геном человека.

Биоинформатику часто противопоставляют вычислительной биологии. Но бывают случаи, когда эти направления в некоторых местах считаются синонимами.

Материалы и методы.

Современные проблемы биоинформатики

- Использовать доступные методы селекции для приобретения новых знаний в области биологии
- Разработка новых методов анализа биологических данных
- Создание новых резервов информации и знаний.
- Совершенствование внутренних алгоритмов биоинформатики и разработка новых

Условия для бурного развития биоинформатики сегодня необходимо:

- Развитие компьютерных технологий и появление мощной вычислительной техники, что привело к ускорению процесса обработки данных;
- Ускорение процесса обмена информацией с развитием информационных технологий, появление различных порталов и форумов способствовало быстрому поиску и получению необходимой информации и обмену опытом;
- Совершенствование технологий программирования, появление объектно-ориентированного и визуального программирования привели к созданию большого количества программных приложений в области биотехнологии.
- Увеличение и уточнение биологических данных, подлежащих обработке в области биотехнологии

Сегодня, в зависимости от объектов интереса и используемого для этого программного обеспечения, основные направления деятельности в области биоинформатики осуществляются по следующим направлениям:

- анализ последовательности (биоинформатика последовательности);

-пространственное картографирование (систематическая биоинформатика);

- вычислительная геномика.

Работа в этих областях требует высокой скорости и точности обработки информации. Некоторое программное обеспечение, используемое в биоинформатике:

JalView — редактор для множественного выравнивания последовательностей нуклеотидов и аминокислот;

Популяции — популяционно-генетический анализ;

МЕГА — молекулярно-эволюционный генетический анализ;

Mesquite — программа для сравнительной биологии на Java.

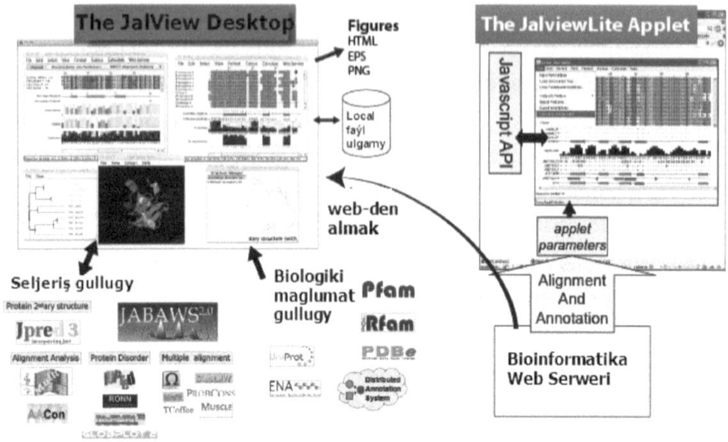

Рисунок-1. Илюстарция работа программы **JALVIEW**

JalviewLite — это веб-версия Jalview, которая используется на многих веб-сайтах.

Jalview Desktop — это программный инструмент для биологов и биоинформатиков, позволяющий создавать, управлять и анализировать несколько последовательностей.

Ускорение алгоритмов сортировки. За время развития биотехнологии было собрано огромное количество биологических данных. То есть есть миллионы генетических тестов, обработанных программой. Один документ, содержащий последовательности нуклеотидов, содержит более ста миллиардов символов. Здесь найти нужный ген так же сложно, как найти нужную цитату в большой библиотеке. Конечно, для обработки таких данных необходима высокая скорость.

Развитие технологий сделало возможным создание машин с высокой скоростью вычислений. Но скорость вычислений зависит не только от железа, но и от скорости программного обеспечения. Разные языки программирования, разные версии каждого языка программирования предлагают разную скорость обработки данных.

Заключение. Программное обеспечение для биоинформатики является важным инструментом для решения задач биоинформатики. Увеличение скорости итерационных операций по анализу последовательностей в некоторой степени повысит эффективность процесса анализа. Однако интеллектуальный подход к разработке алгоритма программы, за счет увеличения искусственного интеллекта программы, позволяет выйти на совершенно новый этап в этой области, и можно поднять биоинформатику на новый уровень.

Список литературы

1. Стюарт Рассел, Питер Норвиг. Искусственный интеллект: современный подход (AIMA), 2-е издание.: Пер. с англ. – М.: Издательский дом «Вильямс», 2006, 1424 с.

2. А.Н.Огурцов. Основы биоинформатики. Учебн.пособие по курсам «Биоинформатика и информационная биотехнология» и «Биоинформатика и фармокоинформатика». Харьков НТУ. 2013. 399 с.

3. Чуриев М.М. Некоторые проблемы защиты компьютерных программ от отладчиков-дебаггеров // Аспирант и соискатель. ISSN 1608-9014, Издательство «Спутник +», Москва – 2013. – №1. – с. 64-68.

УДК 004

СОЗДАНИЕ ПРОГРАММНОЙ СИСТЕМЫ, АВТОМАТИЗИРУЮЩЕЙ РАБОТУ РИЭЛТОРОВ

М.А. Атаев, П.С. Мырадов, Дж.Р. Абдыев
Государственный энергетический институт Туркменистана
Мары, Туркменистан

Аннотация. Это программное обеспечение называется «Создание программной системы, автоматизирующей работу риэлторов». Для реализации программы был выбран язык программирования Embarcadero Delphi XE7. Эта программа планирует дифференцировать работу риелторов, отказавшись от бумаги, блокнотов, ручного поиска, организовав ежедневные и еженедельные задачи и заключать контракты с клиентами, и все это можно сделать быстро. Также будет возможность просматривать и сравнивать статистику выполненных мероприятий по годам, месяцам, неделям.

Ключевые слова: программное обеспечение, автоматизация работы, риэлторы.

Разрабатываемое программное обеспечение состоит из семи разделов. Эти разделы: «Работа и задания», «Мои объекты», «Выбор объектов», «Договоры», «Клиенты»; «Статистика», «Об авторе».

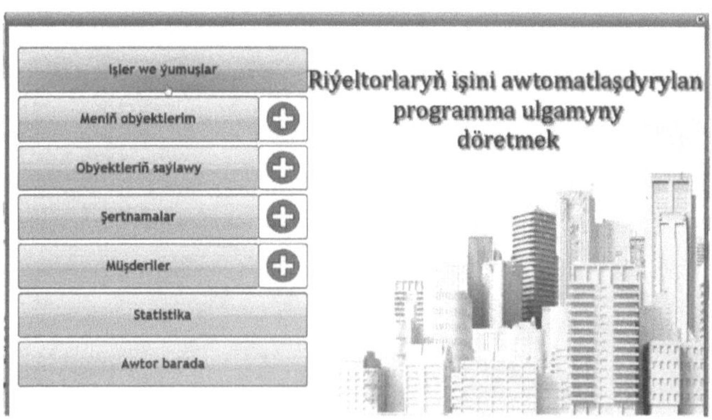

Рис. 1. Интерфейс главного окна программы

Главное окно этой программы имеет 7 основных разделов. При выборе первого раздела становится доступным раздел «Работа и задания». В этом разделе вы можете записывать время и дату выполнения задач.

Второй раздел — «Мои объекты». В этом разделе отображается список объектов, принадлежащих риелтору. Рядом с этим разделом также размещена кнопка «Добавить объект». Для добавления объектов здесь предусмотрены поля.

В третьем разделе «Выбор объектов» можно вводить объекты, которые имеют определенные предпочтения клиента. Здесь также создан дополнительный раздел для добавления новой выборки объектов, т.е. если вам нужно добавить объекты в непредвиденных условиях, вы можете добавить их в поле дополнительных заметок.

В четвёртом разделе «Договоры» отображается список договоров между клиентом и риелтором. Создан раздел для заполнения соответствующей формы добавления договора.

В пятом разделе «Клиенты» можно зарегистрировать любого клиента с запросом на поиск объекта. Для этого нужно нажать на кнопку + рядом с кнопкой «Клиенты». Его можно отобразить, если с ним связан готовый объект. Если нет, то его можно оставить пустым.

В шестом разделе «Статистика» содержится отчет о деятельности риелтора за определенный период времени (разные месяцы, выбранный день, неделя, 1 год...).

Седьмой раздел называется «Об авторе» и в этом разделе содержится информация о разработчиках программы.

В разделе «Работа и задания» вы можете зарегистрировать список задач, которые необходимо выполнить, с указанием времени и даты. Для выполнения этих задач в календаре выбирается соответствующая дата и указывается время. Затем нажимается кнопка «Зарегистрировать задание». Соответствующие задания зарегистрированы. Нажмите кнопку «Просмотреть список заданий», чтобы просмотреть зарегистрированные задания.

Рис. 2. Интерфейс раздела «Работа и задания»

```
procedure TForm2.BitBtn2Click(Sender: TObject);
begin
form3.Show;
form3.ADOQuery1.Close;
form3.ADOQuery1.SQL.Clear;
form3.ADOQuery1.SQL.Text:='select wagty, senesi,bellik from yumush where Yerineyetirilisi="Ýerine ýetirildi" ';
form3.ADOQuery1.Open;
form3.adoquery1.Active:=true;
end;
```

При нажатии на кнопку соответствующего раздела «Мои объекты», отображается список объектов, принадлежащих риелтору. Здесь вы можете выполнять различные операции фильтрации над объектом. Например, следующий фильтр применяется к местоположению этажа объекта. Строку фильтра можно скрыть, нажав кнопку «Скрыть фильтр», а строку фильтра можно снова отобразить, нажав кнопку «Показать фильтр».

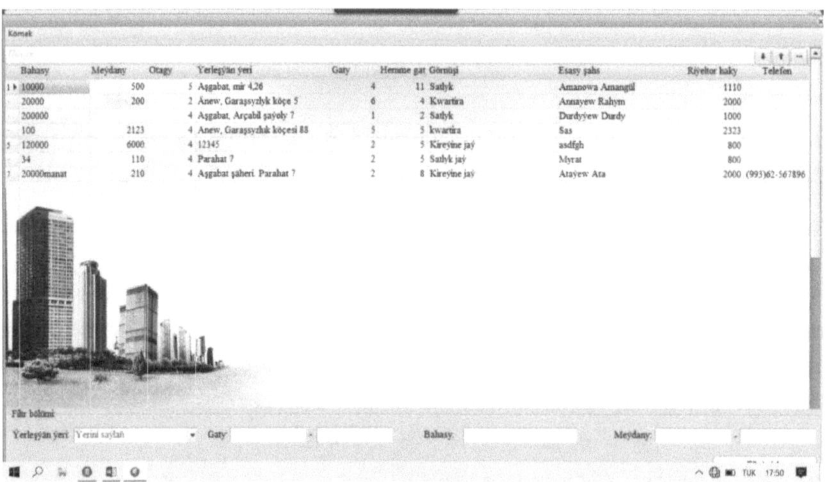

Рис. 3. Интерфейс раздела «Мои объекты»

В разделе «Мои объекты» для добавления новых объектов предусмотрена кнопка. Для выбора этого раздела необходимо нажать кнопку «+» в программе. В результате откроется окно. Вы должны заполнить поля, чтобы добавить объект. Добавленный объект можно просмотреть, нажав кнопку «Мои объекты».

«Выбор объекта». При выборе этого раздела вводятся объекты с конкретным выбором конкретного заказчика, то есть, если необходимо найти объект для данного заказчика только по отображаемым параметрам, необходимо заполнить поля для регистрации. Например, необходимо найти дом, соответствующий параметрам, указанным для Дурдыевой Сахры. В этом случае он должен заполнить эту страницу.

Чтобы добавить новое выделение к выборке объектов, необходимо сделать следующее. В этом окне можно ввести различные параметры. Если заказчику нужен объект с условиями, не предусмотренными в программе, он может написать эти условия в поле дополнительных примечаний. Также возможен поиск в окне списка объектов. Для выполнения этой операции достаточно ввести соответствующее слово в строку «Поиск».

В разделе «Договоры» отображается список договоров между клиентом и риелтором. Здесь есть возможность передать договор в редактор MS Word и подготовить договор автоматически. После нажатия на кнопку подготовить договор в редакторе Ms Word, в появившемся окне напишите имя Риэлтора и условия, которые должны быть указаны в договоре. Этот объект можно удалить из списка, нажав кнопку «Удалить проданный объект».

Нажатие кнопки «Учёт» - означает, что объект продан риелтором по договору. Поэтому риелтор должен нажать кнопку «Учёт» после подписания договора. Он будет автоматически перенесен в раздел статистики.

Для добавления договора необходимо также заполнить соответствующую форму. Здесь после того, как в окне объектов появится список объектов, имеющихся в резерве риелтора, можно выбрать один из этих объектов.

При выборе раздела «Клиенты» можно зарегистрировать любого клиента, обратившегося с запросом на поиск объекта. Для выполнения этой операции нажмите на кнопку «+», т.е. «Добавить», расположенную рядом с разделом «Клиенты». Если для него есть готовый объект, его можно увидеть, если нет, то его можно оставить пустым.

Если у этого покупателя есть товары по душе, и вам нужна информация о нем, то нажмите кнопку «Отправить сообщение». Выберите соответствующий объект из появившегося на экране окна, и параметры выбранного объекта будут автоматически записаны. Также необходимо ввести имя пользователя, пароль и адрес электронной почты. Адрес электронной почты клиента отображается автоматически. Вы также можете удалить эту запись клиента, нажав кнопку «Удалить».

Раздел статистики содержит отчет о деятельности риелтора за определенный период времени (по месяцам, выбранному дню, неделе, 1 году...). Эти отчеты можно импортировать в указанные редакторы, нажимая кнопки MS Word и MS Excel.

В заключение можно сказать что данное программное обеспечение разработана для автоматизации работ риелторов и самое главное отличие от похожих многих программ является, что интерфейс программы понятном для заказчиков на туркменском языке.

Список литературы

1. Умный щит. Щиты распределения электроэнергии с цифровой поддержкой. //Schneider Electric, 2015.
2. Чуриев М.М. Вопросы разработки современных методов защиты от троянских программ с использованием системного реестра // Сборник трудов международной научно-практической конференции студентов, магистрантов и аспирантов «рынок и эффективность производства-13». – Көкшетау, Казахстан: 2016.– с. 535-537.
3. Стюарт Рассел, Питер Норвиг. Искусственный интеллект: современный подход (AIMA), 2-е издание: Пер. с англ. – М.: Издательский дом «Вильямс», 2006, 1424с.
4. S. Nazarov, B. Jumayev. Programming-based laboratory assignments for undergraduate students of power engineering fields. Compsystech 2021. Ruse, Bulgaria. DOI: 10.1145/3472410.347429.

УДК 004

ПРЕДОСТАВЛЕНИЕ ИНФОРМАЦИОННОГО ПРОГРАММНОГО ОБЕСПЕЧЕНИЯ О МУЗЕЯХ СТРАНЫ

М.А. Атаев, П.С. Мырадов, Дж.Р. Абдыев
Государственный энергетический институт Туркменистана
Мары, Туркменистан

Аннотация. Это национальное программное обеспечение называется “Создание информационной программы о музеях Туркменистана на языке C#”. Для разработки программного обеспечения сначала была создана база данных, а затем разработан программный продукт, работающий с этой базой данных.

Чтобы запустить Visual Studio 2012, выберите “Пуск” “Все программы Microsoft Visual Studio 2012 Visual Studio 2012, дважды щёлкните по нему или нажмите клавишу Enter. После запуска появится следующее окно.

Это национальное программное обеспечение называется “Создание информационной программы о музеях Туркменистана на языке C#”. Для разработки программного обеспечения сначала была создана база данных, а затем разработан программный продукт, работающий с этой базой данных. На изображении выше показано главное окно этого программного продукта. С помощью этой программы пользователи могут просматривать информацию о различных музеях нашей страны. Программа отображает следующую информацию о каждом музее:

Название музея

Адрес местонахождения

Номера телефонов

Фотографии

При разработке программы были использованы следующие компоненты:

- Label,
- GroupBox,
- Button,
- PictureBox,
- DataSet,
- DataGridView.

При использовании каждого компонента его свойства и события были подробно изучены отдельно.

На экране появляется такое окно. В этом языке программирования компоненты размещаются в контексте ToolBox. Чтобы увидеть его, запустим новый проект, нажав на пункт «Новый проект». После этого нам нужно запустить форму Windows Forms Applications. Она размещена на изображении ниже. После этого открывается окно. Ниже показано состояние после открытия новой формы. После открытия формы мы размещаем на ней необходимые компоненты. Сначала мы размещаем на форме компонент PictureBox, чтобы разместить на ней файл изображения с названием программы. Затем мы загружаем изображение на форму, используя свойство Image компонента PictureBox. Затем мы размещаем на форме 3 компонента GroupBox. Внутри каждого компонента GroupBox мы размещаем компоненты PictureBox и Label. Затем мы размещаем на форме компонент DataGridView для подключения нашей программы к базе данных. Этот компонент позволяет нам работать с отдельными таблицами в базе данных. Мы устанавливаем его свойство Visible в значение false, чтобы сделать этот компонент невидимым при запуске программы. Как видно, мы можем отображать информацию только о 3 музеях одновременно, но теперь мы размещаем на форме два компонента Button для отображения информации о других музеях. В результате при открытии программы будут отображены первые три музея из базы данных программы. При нажатии кнопки "Следующая страница" будут отображены вторые три музея из базы данных. При повторном нажатии кнопки "Предыдущая страница" будут отображены оставшиеся три музея. Теперь мы поместим на нашу форму компонент MenuStrip, который позволит переходить к другим разделам приложения.

Наконец, интерфейс нашей программы готов, и в следующем разделе будут показаны коды, которые мы использовали в нашей программе.

Заключение: В результате разработки компьютерной программы «Museums Information Provider» пользователи получили возможность получать цифровую информацию о музеях нашей страны, видеть их местоположение, телефоны и фотографии. Программа работает на основе базы данных и отображает информацию в удобном для пользователя интерфейсе. Данный проект служит популяризации культурного наследия нашей страны посредством цифровой системы, систематическому хранению информации и обеспечению её доступности для широкой публики. Таким образом, программа стала важным национальным программным обеспечением, способствующим цифровому развитию музейной системы нашей страны.При создании программы для связи с информационными ресурсами использовались язык программирования C# и среда Visual Studio 2012. Пользовательский интерфейс разработан понятным, простым в использовании и эстетически привлекательным. Такие возможности позволяют каждому, кто работает с программой, легко находить музейную информацию, просматривать изображения и быстро получать необходимую информацию.

Данная программа не только отображает информацию, но и служит полезным ресурсом для образовательных и научных исследований. В дальнейшем, путем усовершенствования программы, возможно ее развитие в версии, которая автоматически обновляет музейную информацию, использует службы геолокации (GPS) и работает на нескольких языках.

Таким образом, данная работа является важной частью цифровой трансформации музейной системы нашей страны и имеет большое значение для поддержания информации в современном виде, повышения уровня знаний общественности и сохранения культурного наследия. Внедрение такого программного обеспечения позволяет развивать информационные технологии во взаимодействии с культурной системой.

Список литературы

1. М. Атаев, М. Чуриев, Г. Атаев, А. Агаев и Дж. Абдиев. "Развитие технологий создания программного обеспечения в электроэнергетической отрасли Туркменистана". 27 марта 2023 г. DOI:10.1088/1755-1315/1154/1/012063.

2. Фелдер Р., Брент Р. Преподавание и изучение STEM: практическое применение. – Сан-Франциско: Jossey-Bass, 2016.

3. С. Назаров, М. Гурбансахедов. Учебник «Основы теории кибербезопасности». Ашхабад, 2023.

4. Атаманов Б.Ю., Чуриев М.М. Некоторые вопросы кибербезопасности. Научный журнал "Математические методы в технологиях и технике", №10, 2022., © ООО "Сарлен-Алекс", 2022, ISSN 2712-8873, DOI 10.52348/2712-8873_MMTT_2022_10_115, стр. 115-122.

УДК 004

ПРОГРАММНОЕ ОБЕСПЕЧЕНИЕ ДЛЯ ГРУЗОПЕРЕВОЗОК

М.А. Атаев, П.С. Мырадов, Дж.Р. Абдыев
Государственный энергетический институт Туркменистана
Мары, Туркменистан

Аннотация. Программа работает на основе понятной и понятной панели управления, обеспечивающей безопасный доступ. Интерфейс входа в систему обеспечивает безопасный доступ по имени пользователя и паролю, а также возможность создания новой учётной записи при необходимости. Это гарантирует доступ к системе только авторизованным пользователям и защиту конфиденциальной информации, связанной с перемещением товаров.

Программа работает на основе понятной и понятной панели управления, обеспечивающей безопасный доступ. Интерфейс входа в систему обеспечивает безопасный доступ по имени пользователя и паролю, а также возможность создания новой учётной записи при необходимости. Это гарантирует доступ к системе только авторизованным пользователям и защиту конфиденциальной информации, связанной с перемещением товаров. На странице регистрации запрашивается необходимая информация, такая как адрес электронной почты и пароль, а также предоставляются инструкции по созданию надёжного пароля.

Раздел “Товары” позволяет фильтровать товары по релевантной информации. Пользователи могут ввести название товара, начальную и конечную цену, а также соответствующую категорию (например, “Все” или “В наличии”) и использовать кнопки «Поиск и фильтр» и “Экспорт в CSV” для поиска или экспорта данных. Эта функция очень полезна специалистам по логистике для быстрого поиска необходимых товаров и подготовки отчётов.

Раздел «Категории» предназначен для организации товаров. Здесь отображаются такие категории, как “Технологии”, и их описания. Категоризация важна для общей организации товаров и упрощает управление различными типами товаров.

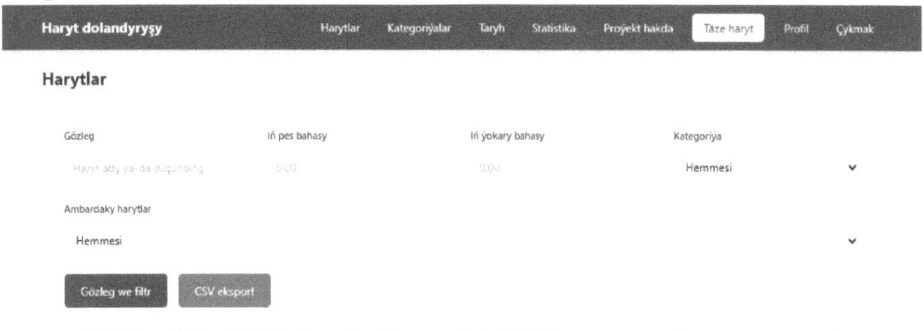

В разделе “Профиль” отображаются имя, фамилия и адрес электронной почты пользователя. Здесь также отображается информация о его активности, например, о новых добавленных элементах и истории активности. Это своего рода форма подотчётности в системе.

Одной из уникальных особенностей программы является возможность создавать объекты на карте и отслеживать их местоположение. Раздел “Создать объект” позволяет ввести название объекта, его описание, количество, цену, категорию и координаты (широту и долготу). Интерфейс карты, работающий с OpenStreetMap, превращает это местоположение в указатель, вводя координаты или выбирая его непосредственно на карте. Например, на карте отображается объект в Ашхабаде, что визуально указывает на местоположение объекта.

Раздел “История покупок” предоставляет полный обзор движения товаров. Пользователи могут фильтровать историю по категории, ценовому диапазону и другим критериям. Также есть возможность просмотреть карту на весь экран или сохранить её как изображение. На карте отображается маршрут, указывающий направление движения товаров.

Кроме того, система предусматривает административный модуль, предназначенный для управления пользователями, их ролями и правами доступа. Администратор может добавлять новых пользователей, редактировать их данные, блокировать учетные записи при нарушении правил безопасности и отслеживать действия в системе. Это позволяет поддерживать высокий уровень контроля и прозрачности всех процессов.

Программа интегрируется с внешними сервисами и платформами, что делает возможным автоматический обмен данными между различными отделами компании.

Например, данные о движении товаров могут быть синхронизированы с бухгалтерской системой, системой складского учёта или CRM-платформой. Такая интеграция способствует ускорению документооборота и снижает количество ручных операций.

Для повышения удобства и безопасности предусмотрена многоуровневая система авторизации, включающая использование одноразовых кодов (OTP) или двухфакторной аутентификации (2FA). Это значительно снижает риск несанкционированного доступа и защищает коммерческие данные компании.

Программа имеет адаптивный дизайн, что позволяет работать с ней как на стационарных компьютерах, так и на мобильных устройствах. Это особенно важно для логистических менеджеров и водителей, которые могут оперативно вносить данные о перемещении грузов прямо с планшета или смартфона.

Дополнительные инструменты аналитики позволяют формировать интерактивные отчёты и графики, демонстрирующие динамику перевозок, загрузку транспортных средств, среднее время доставки и другие ключевые показатели эффективности (KPI). Эти данные помогают руководству принимать обоснованные решения, оптимизировать маршруты и снижать издержки.

Также предусмотрен модуль уведомлений и напоминаний, который информирует пользователей о критически важных событиях — например, о прибытии товара, необходимости обновления данных или истечении срока действия контракта.

В перспективе развитие программы может включать внедрение искусственного интеллекта (AI) для прогнозирования задержек поставок, оптимизации маршрутов и автоматического анализа логистических рисков. Это сделает систему ещё более интеллектуальной и эффективной для комплексного управления товаропотоками.

Программа также позволяет использовать различные уровни доступа для пользователей. Например, логист может видеть только данные о маршрутах и транспортных средствах, тогда как администратор имеет полный доступ к статистике, отчётам и настройкам. Это помогает соблюдать внутреннюю иерархию и предотвращает несанкционированное редактирование данных.

Одним из ключевых преимуществ является наличие гибкой системы отчетности. Пользователи могут самостоятельно настраивать параметры отчётов, выбирая нужные поля, временные интервалы и тип визуализации (таблицы, графики, диаграммы). Готовые отчёты можно экспортировать в форматы CSV, PDF или Excel для дальнейшего анализа.

Для улучшения пользовательского опыта реализован интерактивный интерфейс карты, который позволяет не только отслеживать местоположение объектов, но и анализировать логистические маршруты. Например, пользователь может просмотреть, какой маршрут является наиболее коротким и экономичным с учётом пробок, расстояний и времени в пути.

Также рассматривается возможность интеграции с системами машинного зрения (Computer Vision) для автоматического распознавания номеров транспортных средств и контроля погрузочно-разгрузочных операций с помощью камер наблюдения.

В целом, данная веб-программа представляет собой современное и инновационное решение для цифровизации логистических процессов. Она сочетает в себе простоту использования, высокий уровень безопасности и мощные аналитические возможности, что делает её незаменимым инструментом для компаний, стремящихся к автоматизации и повышению эффективности управления грузопотоками.

Заключение: В результате проведённого анализа можно сделать вывод, что внедрение системы «Цифровой контроль за перемещением грузов» играет ключевую роль в модернизации логистических процессов. Программа обеспечивает прозрачность, безопасность и оперативность в управлении движением товаров, что особенно важно в условиях растущего объёма перевозок и необходимости оптимизации затрат.

Благодаря использованию современных технологий, таких как GPS-мониторинг, интеграция с картографическими сервисами и аналитическими инструментами, система позволяет эффективно контролировать маршруты, отслеживать местоположение грузов в режиме реального времени и предотвращать возможные задержки.

Развитая структура интерфейса, наличие модулей для фильтрации, экспорта и аналитики данных, а также система авторизации пользователей делают программу универсальным инструментом для всех участников логистического процесса.

Таким образом, цифровая система контроля за перемещением грузов способствует повышению эффективности транспортных операций, снижению операционных рисков и улучшению качества обслуживания клиентов. Внедрение подобных решений является важным шагом на пути к цифровой трансформации логистической отрасли и развитию «умных» транспортных систем будущего.

Список литературы

1. М. Атаев, М. Чуриев, Г. Атаев, А. Агаев и Дж. Абдиев. "Развитие технологий создания программного обеспечения в электроэнергетической отрасли Туркменистана". 27 марта 2023 г. DOI:10.1088/1755-1315/1154/1/012063.

2. Фелдер Р., Брент Р. Преподавание и изучение STEM: практическое применение. – Сан-Франциско: Jossey-Bass, 2016.

3. С. Назаров, М. Гурбансахедов. Учебник «Основы теории кибербезопасности». Ашхабад, 2023.

4. Атаманов Б.Ю., Чуриев М.М. Некоторые вопросы кибербезопасности. Научный журнал "Математические методы в технологиях и технике", №10, 2022., © ООО "Сарлен-Алекс", 2022, ISSN 2712-8873, DOI 10.52348/2712-8873_MMTT_2022_10_115, стр. 115-122.

УДК 004

БЕЗОПАСНОСТЬ ПРОГРАММНОГО ОБЕСПЕЧЕНИЯ «КЛИЕНТОВ»

М.А. Атаев, П.С. Мырадов, Дж.Р. Абдыев
Государственный энергетический институт Туркменистана
Мары, Туркменистан

Аннотация. В перспективе дальнейшее развитие программы, включая онлайн-платежи, уведомления и мобильную версию, позволит сделать систему ещё более удобной, многофункциональной и адаптированной под современные требования бизнеса. Таким образом, веб-программа служит эффективным инструментом для цифровизации процессов управления заказами и повышения конкурентоспособности компании.

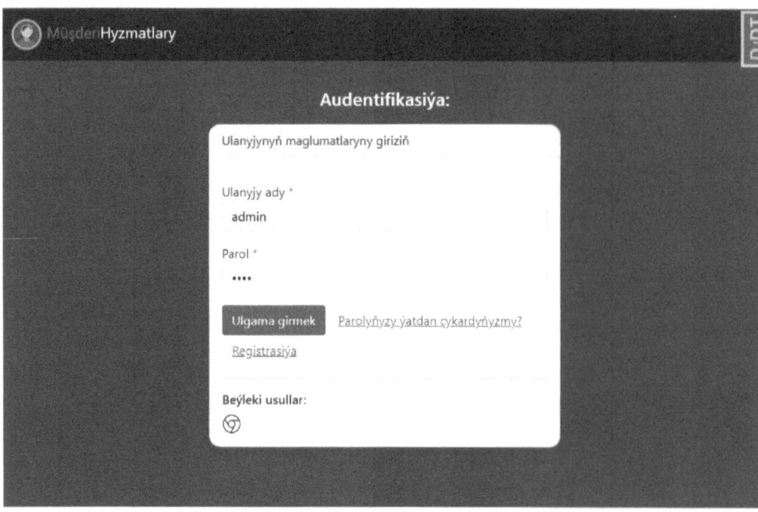

Авторизация (вход):

• Эта страница предназначена для входа пользователей в систему. Войти в систему можно, заполнив поля “Имя пользователя” и “Пароль” и нажав кнопку “Войти”.

• Если вы забыли пароль, вы можете перейти по ссылке “Забыли пароль?”, чтобы войти в систему, используя свой адрес электронной почты, и создать новый пароль.

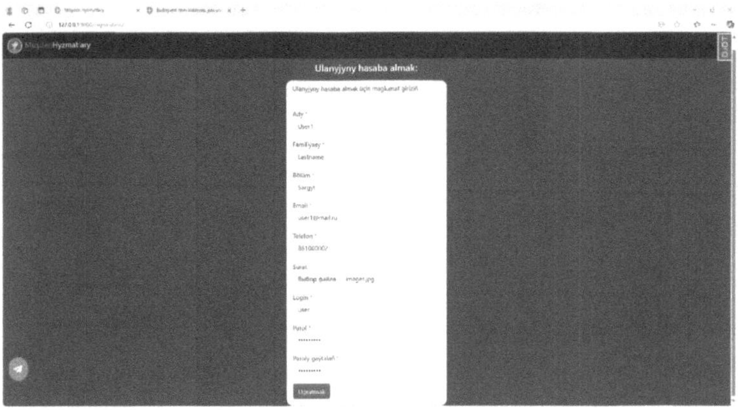

Чтобы получить учетную запись пользователя:
Страница ввода информации для регистрации нового пользователя. Заполните поля «Имя», «Фамилия», “Заказ”, “Email”, “Телефон”, “Фото” (личное фото), “Логин”, “Пароль” и “Повторите пароль” и нажмите кнопку “Отправить” .

После входа в систему открывается главная страница программы. В верхней части главной страницы находится строка меню, в которой отображаются адреса. В данном примере, если нажать на адрес “Вы: admin” (имя пользователя admin), откроется модальное меню.
Это модальное меню содержит следующие ссылки: “Создать задачу”, “Администрирование”, “Личный кабинет”, “Гарантия” и “Выйти”.
При нажатии на ссылку “Создать задачу” мы переходим на следующую страницу — “Создать заказ”.

На этой странице введите название заказа и нажмите кнопку «Далее», в результате чего будет создан новый заказ.

В разделе аналитики программы статистика всех заказов отображается в виде диаграммы и таблицы.

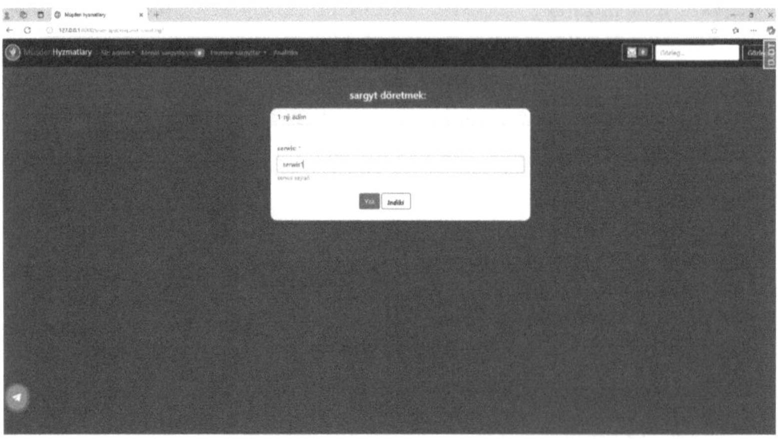

Это приложение предназначено для всех пользователей. Оно позволяет простым пользователям создавать различные заказы, администратор управляет всем приложением, а сотрудники — процессом выполнения заказов.

Конечно! Вот перевод вашего текста на русский язык с сохранением структуры и смысла. Программа обеспечивает учёт заказов клиентов и автоматизирует все процессы, связанные с их обработкой. Каждый пользователь может войти в свой личный кабинет и отслеживать статус заказов, этапы их выполнения и прогресс обработки. Таким образом, программа облегчает взаимодействие между клиентом и сотрудником и делает управление заказами более прозрачным и эффективным. Через админ-панель осуществляется централизованное управление системой. Администратор может добавлять или удалять пользователей, подтверждать заказы, а также обеспечивать безопасность всех данных программы. Такая структура позволяет точно распределять роли и контролировать доступ к информации. Для корректной работы программы используется база данных MySQL, интерфейс разработан с использованием HTML, CSS и JavaScript, а серверная часть может быть реализована на PHP или Python (Django / Flask). Такая архитектура обеспечивает быструю и стабильную работу веб-приложения. В разделе аналитики все заказы отображаются с помощью диаграмм и таблиц. Руководство и сотрудники могут анализировать количество заказов, скорость их выполнения, уровень удовлетворенности клиентов и другие показатели. Это позволяет повысить производительность работы и улучшить качество обслуживания. В будущем веб-программа может быть усовершенствована с добавлением онлайн-платежей, SMS/Email уведомлений, а также мобильной версии. Таким образом, программа не только упрощает текущие рабочие процессы, но и создаёт основу для полного цифрового управления системой заказов. В итоге, веб-программа “Учёт заказов клиентов” является современным решением, использующим возможности современных технологий для повышения качества обслуживания клиентов, оптимизации процессов и повышения эффективности управления бизнесом.

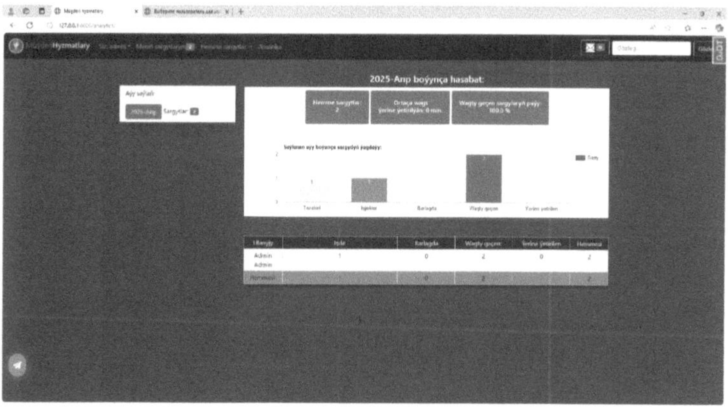

Заключение: Внедрение веб-программы “Учёт заказов клиентов” позволило создать современную и удобную систему для регистрации, обработки и контроля заказов. Программа обеспечивает прозрачность работы, ускоряет взаимодействие между клиентами и сотрудниками, а также упрощает административное управление системой.

Благодаря автоматизации процессов и интеграции с базой данных, пользователи могут легко отслеживать статусы заказов, а руководители — анализировать эффективность работы с помощью диаграмм и таблиц. Такая система повышает производительность работы, снижает вероятность ошибок и улучшает качество обслуживания клиентов. В перспективе дальнейшее развитие программы, включая онлайн-платежи, уведомления и мобильную версию, позволит сделать систему ещё более удобной, многофункциональной и адаптированной под современные требования бизнеса. Таким образом, веб-программа служит эффективным инструментом для цифровизации процессов управления заказами и повышения конкурентоспособности компании.

Список литературы

1. М. Атаев, М. Чуриев, Г. Атаев, А. Агаев и Дж. Абдиев. "Развитие технологий создания программного обеспечения в электроэнергетической отрасли Туркменистана". 27 марта 2023 г. DOI:10.1088/1755-1315/1154/1/012063.

2. Фелдер Р., Брент Р. Преподавание и изучение STEM: практическое применение. – Сан-Франциско: Jossey-Bass, 2016.

3. С. Назаров, М. Гурбансахедов. Учебник «Основы теории кибербезопасности». Ашхабад, 2023.

4. Атаманов Б.Ю., Чуриев М.М. Некоторые вопросы кибербезопасности. Научный журнал "Математические методы в технологиях и технике", №10, 2022., © ООО "Сарлен-Алекс", 2022, ISSN 2712-8873, DOI 10.52348/2712-8873_MMTT_2022_10_115, стр. 115-122.

УДК 004

БЕЗОПАСНОСТЬ ПРОГРАММНОГО ОБЕСПЕЧЕНИЯ «ГЕОПОРТАЛА»

М.А. Атаев, П.С. Мырадов, Дж.Р. Абдыев
Государственный энергетический институт Туркменистана
Мары, Туркменистан

Аннотация. В результате Geoportal: Значительно повышает доступность информации. Представляя различные тематические слои (дороги, места, здания, водные источники) в наглядном и интерактивном виде, он позволяет пользователям быстро просматривать и вычислять данные. Помогает повысить качество принятия решений. Поскольку платформа позволяет легко выполнять точный географический анализ (измерения, анализ буферных зон и поиск атрибутов), необходимый государственным органам, планировщикам и исследователям.

Приложение открывается со страницы “Вход”. На этой странице пользователи должны ввести имя пользователя и пароль для входа в свою учётную запись. Для новых пользователей также доступна кнопка “Регистрация”. Этот раздел обеспечивает вход в приложение и безопасность данных. Фермеры могут использовать этот раздел для создания учётной записи, публикации и управления информацией, связанной со своей деятельностью.

На странице регистрации пользователи вводят имя пользователя, пароль и его подтверждение. Также указаны требования к надёжности пароля (например, он должен быть не менее 8 символов и содержать как заглавные, так и строчные буквы). Эта страница обеспечивает фермерам безопасный вход в приложение и позволяет каждому пользователю индивидуально и безопасно управлять своей информацией.

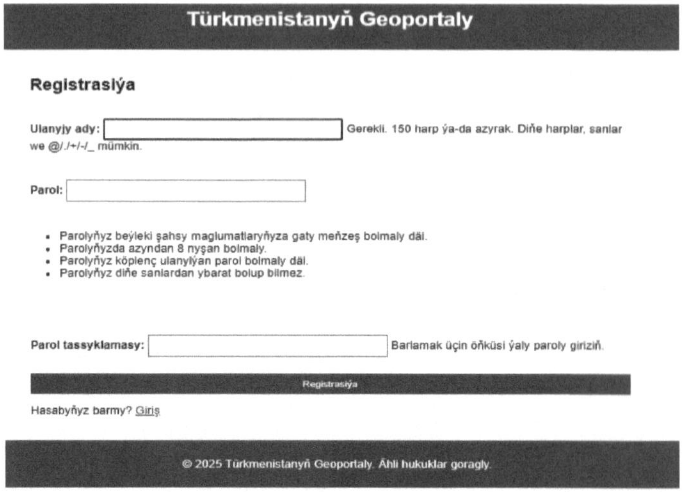

Раздел “Импорт маркеров” предназначен для фермеров, позволяя им размещать маркеры на участках и импортировать данные через CSV-файлы. Здесь пользователи могут добавлять в программу координаты своих участков или данные об урожае, загружая их из файла. Этот раздел помогает собирать и управлять данными из географически удалённых мест в одном месте.

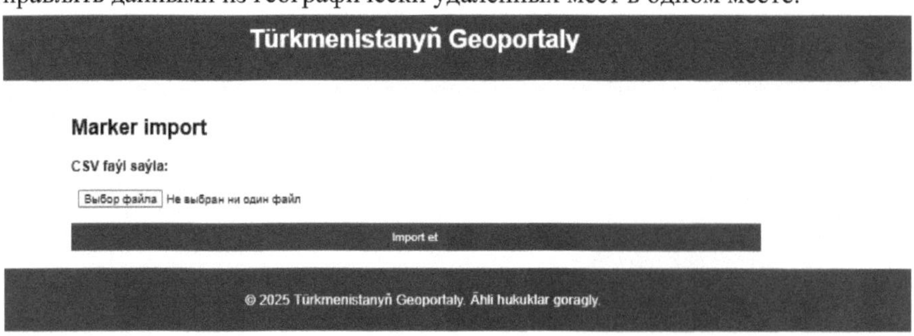

Раздел аналитики отображает статистику для анализа состояния посевов и землепользования. Например, доступны графики “Статистика по посевам” и “Статистика по статусу”. Эти графики позволяют фермерам отслеживать состояние посевов (например, пшеница или отчуждение земли), землепользование и результаты по сезонам. Это позволяет фермерам эффективнее управлять своей деятельностью.

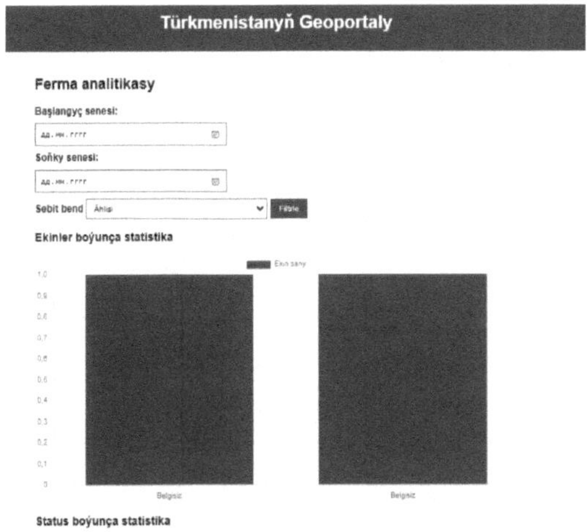

Раздел "Сельскохозяйственная информация" отображает состояние и использование сельскохозяйственных культур по сезонам. Графики содержат информацию о таких культурах, как "Пшеница" или "Землеотвод". Здесь также представлена информация о сроках созревания культур (например, весна или зима) и землепользовании. Этот раздел помогает фермерам принимать необходимые решения для повышения эффективности землепользования.

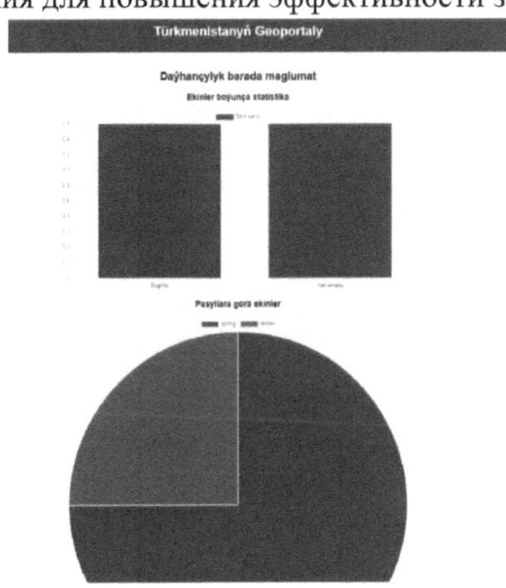

Раздел "Культуры и посевные площади" отображает в таблице периоды выращивания культур и землепользования. Например, для пшеницы указан период "сентябрь-октябрь", а для отвода земель — "февраль-март". В этой таблице фермеры могут легко отслеживать периоды созревания и землепользования своих культур и планировать свою работу. Раздел «Наша карта» отображает географическую карту Туркменистана и позволяет пользователям искать места, устанавливать маркеры или просматривать местоположение. Доступны такие варианты, как “Стандартная карта” и “Спутник”. Фермеры могут использовать этот раздел для удалённого просмотра состояния своих полей и более эффективного управления ими.

Ekinler we ekilýän wagtlary

Ekin	Pasyl	Ekilýän wagty	Ideg usuly	Düşündiriş
Ýer almasy	Ýazlyk	Fewral-Mart	Organiki	Gysgaça maglumat
Käşir	Ýazlyk	Mart-Aprel	Suwaryş	Düşündiriş
Pomidor	Ýazlyk	Mart-Aprel	Organiki	Düşündiriş
Şaly	Ýazlyk	Maý-Iýun	Suwaryş	Suwarmaly
Bugdaý	Gyşlyk	Sentýabr-Oktýabr	Himiki	Düşündiriş

Ideg etmegiň görnüşleri we usullary

- **Suwaryş**: Türkmenistanda daýhançylygyň esasy usuly. Amyderýa we Garagum kanaly ýaly suw çeşmeleri ulanylýar.
- **Organiki daýhançylyk**: Tebigy güýçlendirijiler bilen ekinleriň ýetişdirilmegi, ýer semizligini saklamak üçin.
- **Himiki ideg**: Pestisidler we mineral güýçlendirijiler ulanylýar, ýokary hasyl almak üçin.
- **Tehnologiki usullar**: Häzirki zaman oba hojalyk tehnikalary, dronlar we sensorlar ulanylýar.

Daýhançylyk boýunça maslahatlar

Daýhançylykda ýokary netije almak üçin ýer semizligini yzygiderli barlamak, suwaryşy dogry meýilleşdirmek we möwsümine görä ekin saýlamak möhümdir.

В этом разделе пользователи могут более подробно просмотреть удобное расположение объектов в 3D. Они могут легко найти нужное место, введя название отмеченного места в поле “Обзор”. Также можно проверить статус объекта через опции “Отмеченные места” или “Засаженные места”. Этот раздел упрощает управление географически удалёнными объектами.

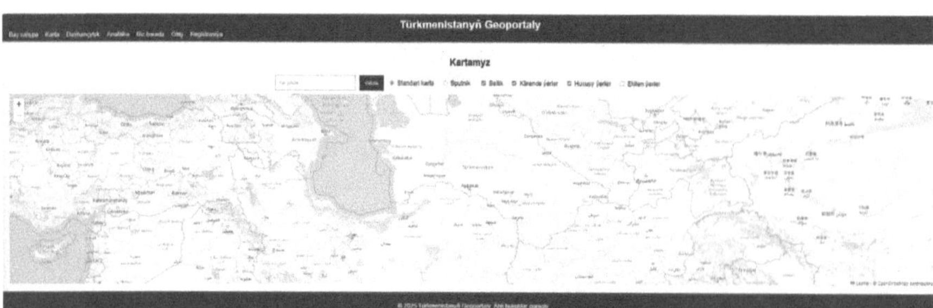

Что такое геопортал?

Геопортал — это веб-интерфейс к географической информационной системе (ГИС). Он предназначен для сбора, просмотра, фильтрации и составления отчётов по пространственным данным из различных источников на единой платформе. Геопортал предлагает более широкий набор функций, чем традиционные карты, и часто создаётся для государственных учреждений, кадастровых служб или академических учреждений.

Будущие возможности:Будущие усовершенствования приложения могут включать следующие области:
3D-визуализация: Отображение зданий и рельефа в трёхмерном (3D) виде.
Развертывание на мобильных устройствах (PWA): Повышение эффективности работы геопортала на мобильных устройствах или в качестве прогрессивного веб-приложения (PWA).
Загрузка данных: Возможность для пользователей загружать собственные пространственные данные (например, шейп-файлы, GeoJSON) и визуализировать их на карте.
Веб-платформа "GEOPORTAL" — важная современная платформа для управления и обмена пространственными данными в цифровую эпоху. Этот проект делает комплексные возможности географической информационной системы (ГИС) доступными всем пользователям через Интернет.
В результате Geoportal:Значительно повышает доступность информации. Представляя различные тематические слои (дороги, места, здания, водные источники) в наглядном и интерактивном виде, он позволяет пользователям быстро просматривать и вычислять данные.
Помогает повысить качество принятия решений. Поскольку платформа позволяет легко выполнять точный географический анализ (измерения, анализ буферных зон и поиск атрибутов), необходимый государственным органам, планировщикам и исследователям. Благодаря соответствию международным стандартам (OGC) он обеспечивает взаимодействие между различными системами и повышает ценность данных. Короче говоря, Geoportal — это не просто веб-сайт, а цифровой мост, который интегрирует, анализирует и предоставляет сообществу геопространственные данные. Это пример более эффективного и открытого способа управления данными.

Список литературы

1. М. Атаев, М. Чуриев, Г. Атаев, А. Агаев и Дж. Абдиев. "Развитие технологий создания программного обеспечения в электроэнергетической отрасли Туркменистана". 27 марта 2023 г. DOI:10.1088/1755-1315/1154/1/012063.
2. Фелдер Р., Брент Р. Преподавание и изучение STEM: практическое применение. – Сан-Франциско: Jossey-Bass, 2016.
3. С. Назаров, М. Гурбансахедов. Учебник «Основы теории кибербезопасности». Ашхабад, 2023.
4. Атаманов Б.Ю., Чуриев М.М. Некоторые вопросы кибербезопасности. Научный журнал "Математические методы в технологиях и технике", №10, 2022., © ООО "Сарлен-Алекс", 2022, ISSN 2712-8873, DOI 10.52348/2712-8873_MMTT_2022_10_115, стр. 115-122.

FSC
www.fsc.org
MIX
Papier | Fördert
gute Waldnutzung
FSC® C083411

Kolibri 360 GmbH
Ferdinand-Jühlke-Straße 7
99095 Erfurt, Deutschland
produktsicherheit@kolibri360.de

Druck:
CPI Druckdienstleistungen GmbH
im Auftrag der
Kolibri 360 GmbH
Ferdinand-Jühlke-Str. 7
99095 Erfurt